Rock Mass Stability Around Underground Excavations in a Mine

Rock Mass Stability Around Underground Excavations in a Mine

A Case Study

Yan Xing

School of Mechanics and Civil Engineering, China University of Mining and Technology, Xuzhou, China, and Department of Mining and Geological Engineering, University of Arizona, Tucson, AZ, USA

Pinnaduwa H.S.W. Kulatilake

School of Resources and Environmental Engineering, Jiangxi University of Science and Technology, Ganzhou, China, and Rock Mass Modeling and Computational Rock Mechanics Laboratories, University of Arizona, Tucson, AZ, USA

Louis Sandbak

Nevada Gold Mines, Turquoise Ridge Joint Venture, NV, USA

CRC Press
Taylor & Francis Group
Boca Raton London New York Leiden

CRC Press is an imprint of the
Taylor & Francis Group, an **informa** business

A BALKEMA BOOK

CRC Press/Balkema is an imprint of the Taylor & Francis Group, an informa business

First issued in paperback 2021

© 2020 Taylor & Francis Group, London, UK

Typeset by Apex CoVantage, LLC

Library of Congress Cataloging-in-Publication Data
Applied for

Published by: CRC Press/Balkema
                Schipholweg 107C, 2316 XC, Leiden, The Netherlands
                e-mail: Pub.NL@taylorandfrancis.com
                www.crcpress.com – www.taylorandfrancis.com

ISBN: 978-0-367-36008-5 (hbk)
ISBN: 978-1-03-208431-2 (pbk)
ISBN: 978-0-429-34323-0 (eBook)

DOI: https://doi.org/10.1201/9780429343230

Contents

Preface

Stability of underground excavations is of great importance to an operating mine because it ensures the safety of working people and operating facilities as well as successful ore production. The rock mass stability around underground excavations heavily depends on the mechanical behavior of intact rock and discontinuities, the discontinuity geometry pattern and the in-situ stress system. In addition, the configurations of excavations and rock supports and the mine construction sequences also play essential roles. Due to the complex geological or in-situ conditions and the engineering constructions, the assessment of rock mass stability for an underground mine could be very challenging and difficult. Some referenced articles and book chapters cover certain aspects of the material covered in this book. However, a textbook tackling of this difficult problem in detail, especially in 3-D, is currently lacking in the rock mechanics literature.

This monograph presents detailed procedures for the stability assessment and support design for an underground mine case study, that is, from the very beginning of collecting geological and geotechnical information, to the exploration of competent analysis tool to incorporate the most relevant factors and to the final part of discontinuum modeling of underground rock mass stability including the evaluation of applied rock supports. These contents are extensively covered in Chapters 2–7. Chapters 1 and 8 provides the introduction and summary of the monograph. The results of performed laboratory tests and numerical analyses that are omitted from the chapters are presented in Appendices A and B for the completion of the study.

This work stems from the dissertation research of the first author, Dr. Yan Xing, at the University of Arizona, under the supervision of Prof. Pinnaduwa H.S.W. Kulatilake. Although some of the results have been published in several journal and conference papers, the monograph contains a large portion of the unpublished materials and is addressed as a comprehensive case study for a practical rock engineering problem.

The authors believe that the monograph is an excellent reference for senior undergraduates, graduate students, researchers and practitioners who work in the underground rock mechanics and rock engineering area in mining engineering, civil geotechnical engineering and distinct element modeling.

Many individuals have contributed to the preparation of this monograph. The first author would like to express her sincerest gratitude to Prof. Pinnaduwa H.S.W. Kulatilake for his advice, support and meticulous editing of this work. Appreciations are also extended to Senior Geotechnical Engineer Louis Sandbak and the mining company for providing the geological and geotechnical data, rock cores and mine technical tours, for providing some references, photographs and other illustrations, and for the review and approval of the manuscript. Srisharan Shreedharan helped to prepare and conduct the laboratory tests. David Streeter, who is the lab manager at the University of Arizona Geomechanics Laboratory, offered technical support and generous help with respect to laboratory testing. Dr. Moe Momayez, Dr. Jinhong Zhang, and Prof. Roy A. Johnson are also appreciated for their comments on the dissertation research work.

Dr. Xing would like to acknowledge the Chinese Scholarship Council and the University of Arizona Graduate College for providing scholarships to conduct the research, and the Graduate Research Assistantships through Research Contract No. 200–2011–39886, which Prof. Pinnaduwa H.S.W. Kulatilake received from the National Institute for Occupational Safety and Health of the Centers for Disease Control and Prevention. Dr. Xing is also grateful for the financial support received from the Natural Science Foundation of Jiangsu Province (BK20180653) and from the China Postdoctoral Science Foundation (2019M652017) for preparing the monograph.

A draft outline of the monograph was sent to a few reviewers around the world. The authors appreciate their review and comments on the book proposal.

Finally, the authors also thank the many individuals and organizations who freely gave permission to reproduce the published material.

Y. Xing
P.H.S.W. Kulatilake
L. A. Sandbak

Abbreviations

3DEC	3-Dimensional Distinct Element Code
AE	acoustic emission
AECL	Atomic Energy of Canada Limited
ASTM	American Society for Testing and Material
BBT	'Better Be There'
BEM	Boundary Element Method
c-p	common plane
CRF	cemented rock fill
DBS	north zone diabase
DDA	Discontinuous Deformation Analysis
DEM	Discrete Element Method
EDZ	excavation damaged zone
FDM	Finite Difference Method
FEM	Finite Element Method
FLAC	Fast Lagrangian Analysis of Continua
FLAC3D	FLAC in 3-Dimensions
FSM	fictitious stress method
FW	footwall
GRC	ground response curve
GSI	geological strength index
HGB	high grade bullion
HW	hanging wall
ISRM	International Society of Rock Mechanics
Ldeep	lower deep
LDP	longitudinal deformation profile
LHD	load-haul-dump
LM, ML	rock types of limestone > mudstone and mudstone > limestone
MBE	Mine-by Experiment
M-C	Mohr-Coulomb
MPBX	multiple point borehole extensometers
N, E, S, W	north, east, south, and west directions
NMD	nodal mixed discretization
NPB	north pillar basalt

NW, SE, NE, SW	intercardinal directions: northwest, southeast, northeast, and southwest
OC2/3, OC5	Ordovician Comus Formation: unit 2/3 and unit 5
PLAXIS	Plasticity Axi-Symmetry
Q	rock mass quality parameter in Q-system
REV	representative elementary volume
RFPA	Realistic Failure Process Analysis
RMR	rock mass rating according to the Geomechanics Classification
RMT	Roberts Mountain Thrust
RQD	rock quality designation
RSR	rock structure rating
SCC	support characteristic curve
Silseds	siliceous sediments
SRF	stress reduction factor
TR	Turquoise Ridge
UCS	uniaxial compressive strength
UDEC	Universal Distinct Element Code
URL	Underground Research Laboratory
V-Dike	vertical dike
WSM	world stress map

Symbols

Symbols are defined where they are introduced. The used symbols are listed as below:

$\bar{\dot{e}}$	average volumetric strain rate
$F_c^t(\Delta F_c^t)$	axial force (increment) of the reinforcement in cable structural model
$\dot{e}_{ij}, \dot{e}$	deviatoric and volumetric strain rates
$\sigma_{t_{max}}$	maximum tensile strength
$\Delta\varepsilon_1^{ps}, \Delta\varepsilon_3^{ps}$	plastic shear strain increments
$\Delta\varepsilon_3^{pt}$	plastic tensile strain increment
F_C^S	shear force in the grout
$\dot{\varepsilon}_{ij}$	strain rate
$\Delta\varepsilon_m^{ps}$	volumetric plastic shear strain increment
λ	stress relaxation factor
λ_1, λ_2	stress release coefficients used for the installation of tunnel lining and tunnel invert
σ_{3max}	maximum confining stress within which the fitting relation between the Hoek-Brown and the Mohr-Coulomb criterion is considered
σ_{3n}	a factor related to the maximum confining stress, σ_{3max}
σ_{ci}	uniaxial compressive strength of the intact rock
δ_{ij}	Kronecker delta function
ϕ_j	joint friction angle
$\sigma_n, \tau(\Delta\sigma_n, \Delta\tau)$	joint normal and shear stresses (increments)
ϕ_r, c_r, σ_{tr}	residual friction angle, cohesion and tensile strength of the rock mass
ϕ_{rm}, c_{rm}	the friction angle and cohesion of the rock mass
A_{bolt}	reinforcement cross-sectional area
A_c	area of the contact (sub-contact)
a_n, a_s, e_n, e_s	coefficients in the equations of joint normal and shear stiffnesses in the continuously yielding joint model
B	tunnel width
c, ϕ, σ_t	cohesion, friction angle and tensile strength; can be used for intact rock and rock mass as defined locally
c_j	joint cohesion

D	a factor describing the disturbance degree of the rock mass subjected by blast damage and stress relaxation
D_c	reinforcement diameter
D_j	joint deformation
E, μ	Young's modulus and Poisson's ratio; can be used for intact rock and rock mass as defined locally
E_{bolt}	Young's modulus of the reinforcement
E_i	Young's modulus of the intact rock
e^{ps}, e^{pt} $(\Delta e^{ps}, \Delta e^{pt})$	deviatoric plastic shear and tensile strains (increments)
E_{rm}, μ_{rm}	deformation modulus and Poisson's ratio of the rock mass
F	a factor related to the stress history, current stress state, joint peak and basic friction angle, and the joint roughness parameter
F^n	joint normal force
f_s, f_t	shear and tensile failure functions in Mohr-Coulomb plastic model
G_g	grout shear modulus
H_p	rock load factor
H_t	tunnel height
i	joint dilation angle
J_a	joint alteration number
J_n	joint set number
J_r	joint roughness number
J_v	total number of joints per cubic meter
J_w	joint water reduction factor
k	number of the tetrahedral element
K, G	bulk and shear moduli of the rock mass
K_0	horizontal in-situ stress ratio (lateral stress ratio)
K_{bond}	grout shear stiffness
K_n, K_s	joint normal and shear stiffnesses
L	length of the support segment
m_b, s, a	Hoek-Brown constants of the rock mass
m_i	Hoek-Brown constant of the intact rock
m_n	number of elements surrounding the nth element node
n	node number of the tetrahedral element, and also the normal direction
N_ϕ	a function related to the friction angle in Mohr-Coulomb plastic model
S_{bond}	maximum shear force per length of the grout or bond strength
T	joint tensile strength
t_c	grout annulus thickness
T_{max}, F^s_{max}	maximum tensile and shear forces of joints
u_c, u_m	axial displacements of the reinforcement and of the rock mass
u_n, u_s $(\Delta u_n, \Delta u_s)$	joint normal and shear displacements (increments)

u^t	relative axial displacement of the two nodes of the support segment
V_k	volume of the kth tetrahedral element
V_p, V_s	compression and shear wave velocities
x, y, z, X, Y, Z	right-handed Cartesian coordinates
Z	depth of the rock mass below ground surface
ε_1, ε_3	major and minor principal strains, and also axial and lateral strains
σ, ε	normal stress and the corresponding strain
σ_1, σ_3	major and minor principal stresses, and also axial and confining stresses
σ_v	vertical in-situ stress
τ^p	joint shear strength
τ_{peak}	peak shear strength of the grout
ρ	density of the intact rock sample

Chapter 1

Introduction

1.1 Background and motivation of the study

The underground tunnels, which provide access to the mine area and are used to transport the ore, are essential to underground mine projects. Their stabilities are of great importance because it ensures the safety of the working people and operating equipment as well as successful ore production.

Stability conditions around underground tunnels can be affected by a number of factors, either internally or externally. Basically, the type of intact rock, whether it is competent or soft, determines the capability of the material. As a natural geological material, the rock mass contains pre-existing defects such as fissures, fractures, joints, faults, shear zones, dikes and so forth. These discontinuities could significantly weaken the rock mass strength, increase the rock mass deformability and make the rock mass mechanical behavior complicated. The failure modes of the rock masses are closely related to the geometry pattern and conditions of the discontinuities. As far as the underground situation is concerned, in-situ stress field caused by the overburden strata and the lateral stress system needs to be taken into account. Problems around excavations may arise as the stress exceeds the rock strength or as high differential stresses are encountered (Bhasin and Grimstad, 1996; Hudson, 2001). The design configurations of the excavations, such as orientations, sizes and shapes, are also critical factors (Hoek and Brown, 1980; Bhasin et al., 2006; Wang et al., 2012; Shreedharan and Kulatilake, 2016). For instance, the favorably driven direction of the tunnels is that parallel to the maximum principal stress and that perpendicular to the strike of the discontinuities; the circular tunnel shape is ideal to suffer less stress concentrations and failures. After excavation, the support system has to be installed to help the rock mass support itself. The interactions between the rock mass and the support system depend on the factors including the rock mass properties, the geometry and properties of the supports, the geological condition and so forth. Additionally, the response of rock masses is sensitive to the unloading and loading processes (Cai, 2008; Zhao and Cai, 2010). Different excavation and supporting sequences could result in totally different reactions in the rock masses.

Thus, under different combinations of the aforementioned factors, the assessment of rock mass stability for an underground mine is extremely challenging and difficult. Due to the fact that geological conditions and engineering disturbances may differ from one project to another, a study considering the specific factors to a given site is necessary.

The goal of this study is to investigate the rock mass stability around the tunnels and to suggest support design for an underground mine case study. The mining operation at the mine would last another 20–30 years according to the current producing rate. The development drifts, which are used to extract and transport the ore from the mining area, go through the low rock mass quality area or the major fault zones. To reach different ore zones, the tunnels have to be driven in various directions and at different elevations. The geological conditions at the mine were complicated with many fault zones and poor rock mass conditions. Large deformations and failures of the rock masses took place at the tunnel intersections and weak rock mass area. From the perspective of the mine development, the safety and stability of these tunnels need to be ensured; the application of effective supports for the tunnels is of great importance.

1.2   Structure of the monograph

This monograph is organized in eight chapters. The content of each chapter is summarized as follows.

Chapter 1 is the introduction to this research, which briefly describes the background, the motivation and the monograph outline.

Chapter 2 provides the information of the mine site and the tunnel system, including the geological settings, rock mass conditions, in-situ stresses and the excavation and construction information. Then, the used major ground control methodologies and the performed rehabilitation works at the mine are introduced. The rock mass behavior monitored by the field instrumentations and the problem locations figured out through the field observations are presented.

Chapter 3 focuses on the major methodologies used for the rock mass stability in underground excavations, which are the rock mass classification systems, the analytical methods, the field instrumentations and numerical methods.

Some aspects that need to be addressed in the modeling of rock mass stability around underground excavations are highlighted in Chapter 4.

In Chapter 5, the background and some related theories of the distinct element code are provided.

The experimental procedures and the results of laboratory tests are given in Chapter 6. At the same time, the reasons that caused the variabilities in test results are discussed. To avoid tedious presentation, some of the results are given in Appendix A. Similar arrangements are made for Chapter 7.

Chapter 7 presents the details of numerical simulations, which include the development of the model, the performed stress analyses and the analysis and discussions of the numerical results. Based on the predicted rock mass behavior and the performance of applied rock supports, suggestions for support design are provided.

The last chapter summarizes the monograph; major conclusions and the recommendations for future research are also given.

A comprehensive list of literature is provided at the end of this monograph, in the reference section.

Chapter 2

Description of the site, the tunnel system, used ground control methods and field measurements

2.1 Introduction

The studied mine is an underground mine in the USA and has a remaining life of more than 20 years according to the current producing rate. The deposits are sediment-hosted and are generally controlled by the intersection of mineralized faults and stratigraphic units. The ore bodies are dipping approximately 25–45 degrees with a low rock mass quality in the ore zones. Based on the work performed by other mines and the regional data, an empirical relationship is adopted for the estimation of in-situ stresses. The development drifts, which are used to extract and transport the ore from the mining area, are a complex tunnel system with different heading directions and elevations. The major ground control methods applied at the mine include the rock and cable bolts and the shotcrete and mesh support as well as the cemented rock fill. Rehabilitation works have been performed by installing the stronger and longer Swellex or inflatable bolts, and the shotcrete arch sets. According to the field measurements and observations, the problem areas with records of fast movements are discussed. Details of these aspects are presented in the subsequent sections.

2.2 Geologic settings and rock mass conditions

The stratigraphy of the mine area consists of carbonaceous mudstones and limestone, tuffaceous mudstones and limestone, polylithic megaclastic debris flows, fine-grained debris flows, and basalts, all part of the Cambrian-Ordovician Comus Formation. The Comus Formation is divided into several stratigraphic units in the vicinity of the mine. The units are, from bottom to top:

1   Ordovician Comus Formation – unit 2/3 (OC2/3): laminated mudstone and siltstone with thin-bedded carbonates including silty limestone and calcarenite; locally thin, discontinuous thin-bedded siltstone units. This unit can have more limestone than mudstone. Limestones are not greater than 9.1 m (30 ft) thick.
2   Ordovician Comus Formation – unit 5 (OC5): limestone dominant with thinly laminated to thinly bedded mudstones; also contains most of the fragmental units. This unit can have more mudstone than limestone. It is also a reason why some of the mineralization becomes restricted to structures and is the main ore host.
3   North Pillow Basalt (NPB): green aphanitic basalt containing large pillow structures. It can be massive and contain more calcite than quartz veins, including epidote and

pyrrhotite. The thickness of this unit is approximately 91.4 m (300 ft), and there is minimal thrust faulting.

4   North Zone Diabase (DBS): porphyritic green basalt with plagioclase phenocrysts.

5   Main Dacite Dike: porphyritic dacite dike with plagioclase phenocrysts and a medium grey aphanitic groundmass. When altered, the phenocrysts alter to a white clay and the groundmass becomes tan-white or grey when disseminated pyrite is introduced. Grey pyritic veins become prevalent. Minor realgar is present mostly on fracture surfaces but is not seen everywhere.

6   Vertical Dike (V-Dike): very porphyritic dike with plagioclase phenocrysts in a medium grey to purple groundmass. When altered phenocrysts become white within a light grey to white aphanitic groundmass. This dike usually contains realgar on joint/fracture surfaces. Disseminated pyrite is usually introduced giving the grey coloration. This dike can have sill-like features off the main body.

Above the stratigraphic units are more basalts, mudstones, and cherts which may be part of the Ordovician Valmy Formation, with the thickness ranging from 366 m to 457 m (1200–1500 feet). The deposits are sediment-hosted and are generally controlled by the intersection of mineralized faults and stratigraphic units, which majorly occurs in the OC5 and OC2/3 units. Figure 2.1 shows the diagram of the cross-section with major lithology interpretations, major faulting and the mineralized zone in unit OC5 (the area outlined in the middle bottom part).

    The geologic structure model has been constructed at the mine. It shows that five major fault zones exist at the mine area, which include the Getchell group, the Turquoise Ridge (TR) group, the Roberts Mountain Thrust (RMT) group, the 'Better Be There' (BBT) group and the Giant fault.

    The Getchell fault group consists of the Getchell, the Divide and the Separate fault zones, which can also be seen in Figure 2.1. These fault zones strike roughly N15°W and dip 40–50 degrees to the northeast. The thickness of each may range from 15 m (50 ft) to 61 m (200 ft). The lower Getchell fault group generally hosts lower-grade ore within or adjacent to the fault zones. The Separate fault contains or parallels to the Main Dike mineralized zone (see Figure 2.1), which is associated with high-grade mineralized domains. The TR fault group, which contains the TR, the Crusher, and the Smasher faults, is striking approximately N50°W and dipping deeply to the southwest at 60 degrees. Individually, the fault thickness is about 61 m (200 ft), and the faults are separated from each other by about 305 m (1000 ft). It is likely that these faults control the mineralization responsible for the dispersion of ore away from the Getchell fault group. The RMT fault zone lies above the underground model extents and is believed to form the base of the Valmy Formation locally. This fault is generally horizontal and is offset by the TR fault group. The BBT fault zone strikes parallel to the Getchell fault while dipping about 70 degrees to the west. It consists of multiple fault strands over a thickness of 305 m (1000 ft). The BBT and its multiple fault strands are also a key factor in mineralization. Most of the known mineralization for this underground mine is hosted or adjacent to the lower portion of the BBT fault zone. The Giant Fault is generally east-west striking (N80°W) fault sets, dipping about 30 degrees to the southwest.

    Limited drilling information showed that the local minor discontinuities are distributed with a mean spacing of 3–16 m (10–54 ft) and are filled with materials including gouge, breccia clay, calcite and chlorite. Compared to the whole mine area, the selected study region is in a small area inside the mineralized zone outlined by Figure 2.1. Large-scale

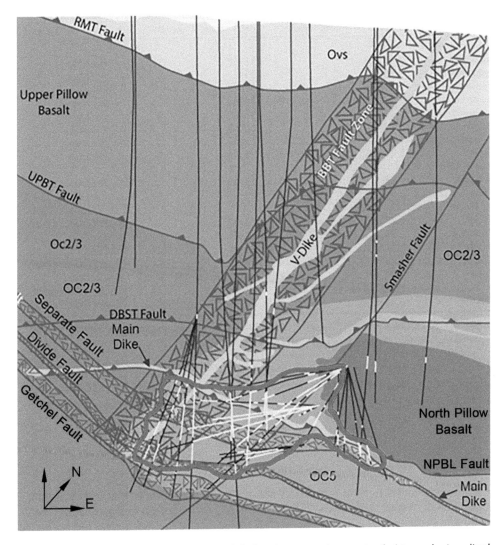

Figure 2.1 Cross section showing major lithology interpretations, major faulting and mineralized zone in middle Comus Formation (OC5)

Source: From Lander (2014).

structures inside the area include the high angle faults striking NW–SE and dipping southwest, the sets that strike NE–SW and dipping northwest and the low-angle faults dipping to the east with 20–40 degrees.

Both geotechnical core logging and geotechnical mapping were carried out to collect the information of underground rock masses. The rock mass conditions were evaluated using the 1976 version of Bieniawski's (1976) RMR system where the input parameters include uniaxial compressive strength (UCS) of rock material, rock quality designation (RQD), the spacing of discontinuities, the condition of discontinuities, groundwater conditions and orientation of discontinuities. Table 2.1 gives the RMR values and class description of the

Table 2.1 RMR values for different geological units at the mine[a]

	Mean RMR (Std. dev.)	RMR range (1 Std. dev.)	Class description
Rock Lithologies			
Megaclastics	44(13)	31–57	Poor-fair
Fragmental	48(13)	35–61	Poor-good
Limestone	51(12)	39–63	Poor-good
Power hill	49(13)	36–62	Poor-good
Npillow basalt	51(12)	39–63	Poor-good
Upillow basalt	43(15)	28–58	Poor-fair
FW sediments	32(10)	22–42	Poor-fair
Silseds	42(14)	28–46	Poor-fair
Intrusive Dikes			
Main dike	33(8)	25–42	Poor-fair
V-dike	40(11)	29–51	Poor-fair
Ore Lenses			
148 W	35(13)	22–48	Poor-fair
148 lower	35(12)	23–47	Poor-fair
FWpond	31(9)	22–40	Poor
Bullion	31(9)	20–40	Very poor-poor
Ldeep	31(10)	21–41	Poor-fair
HGB	31(12)	19–43	Very poor-poor
HWdike	31(10)	21–41	Poor-fair

[a] Data from L. A. Sandbak & A. R. Rai, "Ground support strategies at the Turquoise Ridge Joint Venture, Nevada," *Rock Mechanics and Rock Engineering*, 46 (2013): 437–454.

rock lithologies, intrusive dikes and ore lenses. Generally, the ore bodies and dikes have low RMR values in the range of 20–40; the rock masses outside the ore zone are in better condition, with the average RMR value around 45.

2.3 In-situ stresses

No in-situ stress measurement was carried out at the mine site. Hence, the empirical relation derived by Brown and Hoek (1978), given by Eq. (2.1), is used to determine the vertical in-situ stress (σ_v).

$$\sigma_v = 0.027 \ Z \tag{2.1}$$

where Z is the depth below ground surface.

For the estimate of horizontal in-situ stresses, the equation given by Sheorey (1994) is utilized, where the lateral stress ratio (K_0) is related to the elastic modulus (E_{rm} in GPa) of the rock mass and the depth (Z in m), and is given by

$$K_0 = 0.25 + 7E_{rm}\left(0.001 + \frac{1}{Z}\right) \tag{2.2}$$

Based on the available rock testing results of the elastic modulus and the depth range between 700 m and 1000 m, the lateral stress ratio range of 0.5–1.0 is estimated. This estimation yet is preliminary and more detailed analysis (i.e. sensitivity analysis) should be conducted.

2.4   Layouts and geometry of the tunnel system

The target tunnel system is the development drifts which are utilized to extract and transport the ore from the mining area. Figure 2.2 shows the selected tunnel system. It extends 640 m (2100 ft) along the east-west direction and 305 m (1000 ft) along the south-north direction (see the scale in Figure 2.2a); the elevation of the tunnels, as shown in Figure 2.2b, ranges from 928.7 m to 1125.6 m (3047 ft to 3693 ft). The elevation of the surface is 1615.4 m (5300 ft) so that the depth of the tunnel system is about 489.8–686.7 m. The region inside the square in Figure 2.2a is the selected area of this study which has a history of fast movements according to the field monitoring data. Most of the open drifts are in a horseshoe shape with typical dimensions shown in Figure 2.3a. Backfilling

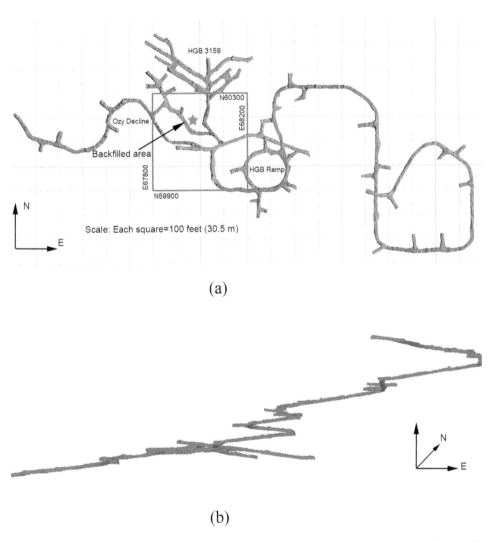

(a)

(b)

Figure 2.2 The overall tunnel system: (a) plan view; (b) elevation view (seeing from the south)
Source: From Xing et al. (2019).

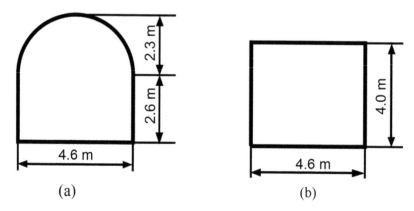

Figure 2.3 Dimensions of the (a) horseshoe tunnels and (b) rectangular tunnels

Source: Reprinted by permission from Springer Nature Customer Service Centre GmbH: Springer, Rock Mechanics and Rock Engineering, "Investigation of rock mass stability investigation around the tunnels in an underground mine in USA using three-dimensional numerical modeling," Y. Xing, P.H.S.W. Kulatilake & L. A. Sandbak, 2018a.

activities have taken place in the northern part of the interested area, as marked in Figure 2.2a. The backfilled area has been excavated in a rectangular shape, given in Figure 2.3b. The tunnels have been driven using jackleg-drilling techniques by the conventional drilling and blasting method or by the underground load-haul-dump (LHD) method. The effect of blasting has been minimized during construction.

2.5   Major ground control methods

2.5.1   Rock and cable bolts

In this underground mine, the primary rock supports including resin bolts, split set bolts and Swellex bolts, and cable bolts have been installed to support the tunnels.

Tables 2.2 and 2.3 provide the specifications for commonly used split sets and Swellex bolts, respectively. These two types of bolts offer friction and mechanical interlocking with the rock masses and have the advantage of easy and quick installation. Therefore, they are widely applied in mining engineering. Due to the possibility of corrosion, the split set bolt is, however, more appropriate to be used as the short-term rather than long-term support (Hoek et al., 1995). Coated bolts are now available to prevent the corrosion for Swellex (see Table 2.3). With connectable bolts by two or three segments together, the length of the Swellex bolts is up to 12 m. For this underground mine, Split sets SS39 and Swellex Pm12 and Pm24 rock bolts were used. Pullout tests in the field showed that the Swellex bolts undertook higher loads than the split sets.

The advantage of resin grouted rebar bolts is that it could provide very high loads. Since the rebar is encapsulated, limited corrosion is another superiority. However, the installation of resin bolts is somehow inconvenient because of the specialized needs of resin setting times or of the unfavorable environment (i.e. high-water areas). In addition, the resins are expensive and have a limited shelf life.

Table 2.2 Specification of split set stabilizer[a]

Description	Bolt type		
	SS-33	SS-39	SS-46
Recommended nominal bit size (mm)	31–33	35–38	41–45
Breaking capacity, average (KN)	107	125	160
Breaking capacity, minimum (KN)	72	89	133
Recommended initial anchorage (KN)	26.5–53	26.5–53	44–80
Tube length (m)	0.9–2.4	0.9–3.0	0.9–3.6
Nominal outer diameter of tube (mm)	33	39	46
Domed plate size (mm^2)	150*150	150*150	150*150
	125*125	125*125	
Galvanised system available	Yes	Yes	Yes
Stainless steel model available	No	Yes	No

[a] Data from Split Set Division, Ingersoll-Rand Company.

Table 2.3 Specifications of typical Swellex rock bolts[a]

Description	Bolt type				
	Pm12	Pm24	Pm24C	Mn12	Mn24
Minimum breaking load (KN)	110	240	200	110	200
Minimum yield load (KN)	100	200	190	90	180
Minimum elongation (%)	10	10	10	20	20
Inflation pressure (MPa)	30	30	30	30	30
Folded tube, diameter (mm)	27.5	36.0	36.0	27.5	36.0
Flat tube, diameter (m)	41.0	54.0	54.0	41.0	54.0
Hole diameter (mm)	32–39	43–52	43–52	32–39	43–53
Bolt length (m)	1.2–6.5	1.8–7.0	4.8–12.3	1.2–6.5	1.8–7.0
Bitumen coated option	Yes	Yes	Yes	Yes	Yes
Plastic coated option	Yes	Yes	Yes	–	–

[a] Data from Atlas Copco (2008).

The cable bolt, which normally has a larger capacity than that of traditional rock bolts, has been developed and widely used in underground mining and civil engineering. Due to the high capacity of cables, the grout quality as well as the cement strength play important roles on the overall performance of the bolt (Hoek et al., 1995).

Three early installations of the rock and cable bolts were applied for the tunnels at the mine, of which the layouts and geometries are presented in Figure 2.4. The first installation, as shown in Figures 2.4a and 2.4b, contains two arrangements. The roof bolts are the 2.44 m resin-grouted rebar bolts; the 1.83 m split set bolts were installed on the ribs. The difference between the two arrangements is the number of roof bolts (3 and 4, respectively). The in-plane spacing and layout of the bolts are shown in the two figures; the two arrangements were placed alternately along the tunnel axis with the out-of-plane spacing of 1.2 m. The second installation includes cable bolts on the roof (Figure 2.4c and 2.4d). Similarly, two arrangements were installed and the out-of-plane spacing is 1.8 m. The third installation is that of the Swellex bolts. Figure 2.4e shows the in-plane configurations of the bolts; 6.10 m long Swellex bolts were installed on the roof and 3.66 m long Swellex bolts on the ribs. The out-of-plane spacing is 1.8 m.

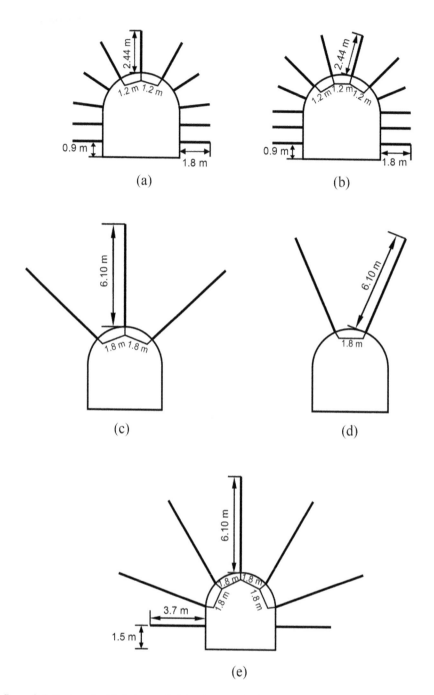

Figure 2.4 Rock and cable bolts applied for the tunnels: (a) first installation – arrangement I (three resin bolts on the roof; eight split set bolts on the ribs); (b) first installation – arrangement 2 (four resin bolts on the roof; eight split set bolts on the ribs); (c) second installation – arrangement I (three cable bolts on the roof); (d) second installation – arrangement 2 (two cable bolts on the roof); (e) third installation (five Swellex bolts on the roof; two Swellex bolts on the ribs)

Source: Reprinted by permission from Springer Nature Customer Service Centre GmbH: Springer, *Rock Mechanics and Rock Engineering*, "Investigation of rock mass stability investigation around the tunnels in an underground mine in USA using three-dimensional numerical modeling," Y. Xing, P.H.S.W. Kulatilake & L. A. Sandbak, 2018a.

2.5.2   Mesh and shotcrete

The mesh and concrete were applied to prevent the unraveling of loose and broken rocks, which may also provide some tensile strength to surrounding rock masses. The primary welded mesh panel used in this mine is the type of 7.6 cm × 7.6 cm (3 in. × 3 in.), with a wire diameter of 6 gauge and in lengths of 2.4 m × 3 m (8 ft × 10 ft). The other available type is the flexible chain link mesh support in galvanized 1.8 m × 7.6 m (6 ft × 25 ft) and 2.4 m × 7.6 m (8 ft × 25 ft) lengths in rolled forms. The overlapped mesh was installed in case the rock would fall out between mats.

Shotcrete is applied to all the primary and long-term development headings, and to the headings where supplementary support to primary bolts and wire mesh is needed. The wet-mix technique is utilized for the mine where the remixed shotcrete mix is delivered to the pump and the accelerator is added at the spray nozzle to quicken the setting time. The average thickness of shotcrete applied to all permanent headings is about 75 mm (3 in.). It is a blend of coarse sand and fine gravel with cement, water, and small amounts of performance additives. The strength of shotcrete is approximately 28–34 MPa (4000–5000 psi) for a 28-day test.

2.5.3   Cemented rock fill

Backfilling is an important part of the ground support for active mining panels at this mine. It could provide lateral support to pillars to stop spalling and prevent collapse, reduce the stope wall closure, create stable mining blocks and perform better than the initial jointed rock masses (Sandbak and Rai, 2013).

The backfilling material is the typically cemented rock fill (CRF), consisting of a mixture of cement and fly ash with crushed and sized aggregate. Test results at the mine showed that the strength and quality of the CRF depend on the aggregate size so that the minimization of fine particles (clay) is the goal to obtain high strength CRF. Figure 2.5 provides the typical aggregate sieve size boundaries used for this mine. The blue and red dashed lines represent the high coarse material with the maximum particle size and the fine material with low backfill strength, respectively; the middle solid line corresponds to the sieve blend with ideal high backfill strength. The targeted minimum backfill strength is 4 MPa (600 psi), while the average strength is generally between 8–10 MPa (1200–1500 psi) for a 28-day strength.

2.5.4   Rehabilitation works

After excavation, rehabilitation works have been performed for the tunnels from time to time, based on the monitored movements of surrounding rock masses. The failed split sets have been replaced by the Swellex or inflatable bolts with stronger resistance; additional rehab was also repeated by installing longer bolts (i.e. the 3.66 m and 6.1 m Swellex bolts).

For the very weak areas and the locations where a permanent rehab solution is needed, the shotcrete arch technique provides robust ground support. The first step of this rehab is to widen the drift profile to the size of approximately 5.0 m (16.5 ft) wide and 5.2 m (17 ft) high for placement of shotcrete arch lattice girders. Then, the floor is sub graded and poured with 0.3 m (1 ft) thickness shotcrete. After that, the arches are erected and attached to the concrete floor. Figure 2.6 shows the schematic cross-section of the shotcrete arch set,

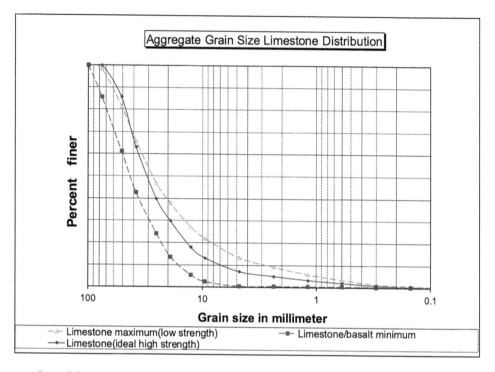

Figure 2.5 Aggregate sieve size optimization curves at the mine

Source: Reprinted by permission from Springer Nature Customer Service Centre GmbH: Springer, *Rock Mechanics and Rock Engineering*, "Ground support strategies at the Turquoise Ridge Joint Venture, Nevada," L. A. Sandbak & A. R. Rai, 2013.

which is consist of four sections of triangular lattice girders that bolted together and is filled with expandable grout in the back and sides. The final completed size of the lattice girder is 4.6 m × 4.9 m (15 ft × 16 ft).

2.6   Field measurements and observations

Tape extensometers and multiple point borehole extensometers (MPBX) were installed at the mine to monitor the tunnel convergence and the movements inside surrounding rock masses, respectively. Figure 2.7 shows the typical tunnel cross section with the instruments. The length of MPBXs is 6.10 m on the roof and is 3.66 m on the ribs. Various anchor points were placed to record the in-rock displacements (see Figure 2.7). The closure rate provides an indication of the rock mass response. In weak rock masses, the convergence rate of more than 1.3 mm per day is likely to cause unstable issues with new cracks observed in the field, whereas the rate of 0.4–0.5 mm per day was considered as a sign of potential failure in stronger rock masses.

The selected area always had rapid movements and was rehabbed several times. As shown in Figure 2.8, tape extensometers (MT17 and MT18) and MPBXs (Sites 1, 2, and 3) were installed near the tunnel intersections. Figure 2.9 provides the photo of the field

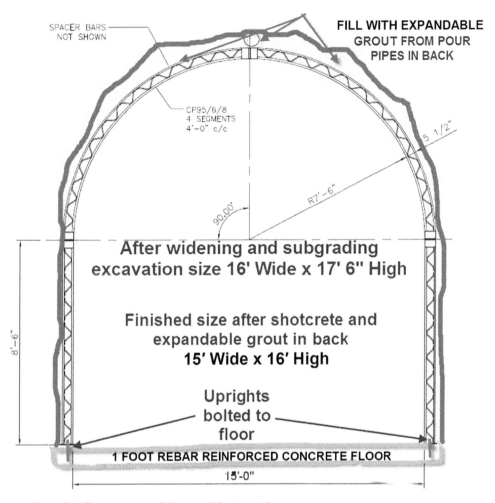

Figure 2.6 Cross section of the typical lattice girder setup

Source: Reprinted by permission from Springer Nature Customer Service Centre GmbH: Springer, *Rock Mechanics and Rock Engineering*, "Ground support strategies at the Turquoise Ridge Joint Venture, Nevada," L. A. Sandbak & A. R. Rai, 2013.

monitoring point MT17, where the mesh and rehabbed shotcrete can also be seen. High tunnel closure rates (0.2–0.3 mm/day) were recorded at locations MT17 and MT18. For the setup of MPBXs, H1 and H2 represent the horizontal instruments on the ribs and V3 is the vertical one on the roof. Monitoring data showed that large displacements (40–60 mm) occurred on the north ribs (H2s) at sites 2 and 3, and that most of the movements were located within 1.2 m from the excavation surface. Slight deformations (1–7 mm) were observed on the roof of the two locations as well as on the ribs of site 1.

Field surveys at the mine site were conducted under the tour given by L. A. Sandbak. Several important locations (i.e. the tunnel intersection and the weak rock mass area) were visited. Figure 2.10 shows the major stops within the selected study area. The corresponding photos are given in Figures 2.11 and 2.12, which were taken by the authors or

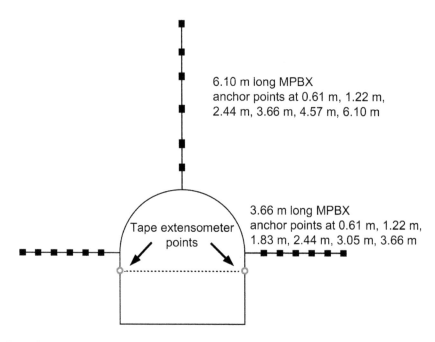

6.10 m long MPBX
anchor points at 0.61 m, 1.22 m,
2.44 m, 3.66 m, 4.57 m, 6.10 m

3.66 m long MPBX
anchor points at 0.61 m, 1.22 m,
1.83 m, 2.44 m, 3.05 m, 3.66 m

Tape extensometer
points

Figure 2.7 Typical tunnel cross section with instruments

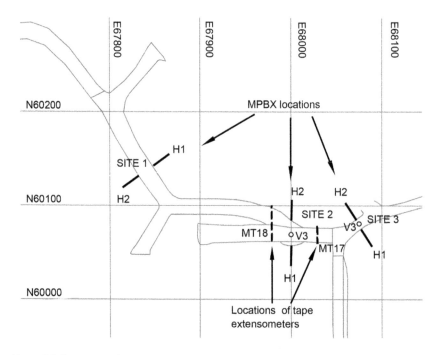

Figure 2.8 Locations of tape extensometers and MPBXs inside the selected study areas (near the tunnel intersections)

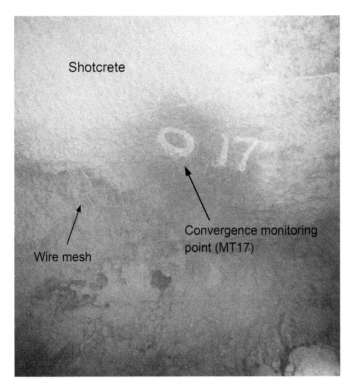

Figure 2.9 Photo of the field monitoring point of MT17

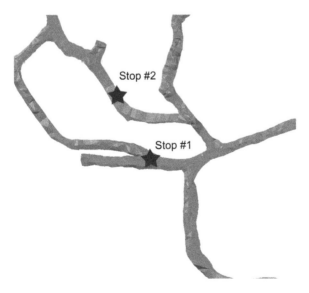

Figure 2.10 Field stops within the selected study areas

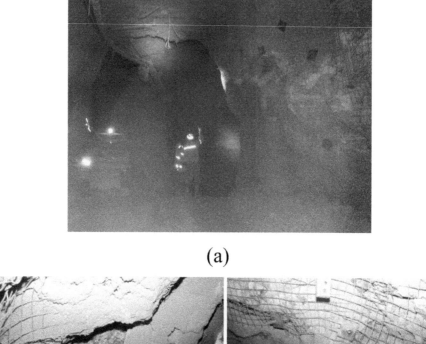

(a)

South wall

North wall

(b)   (c)

Figure 2.11 Photos at Stop #1: (a) overall view; (b) south wall; (c) north wall

Source: From the mining company. From Xing et al. (2016).

provided by the mining company. The overall picture at Stop #1 (Figure 2.11a) presents the highly deformed rock masses, especially the roof of the intersection. With a higher RMR value, the north wall (Figure 2.11c) at Stop #1 had undergone a lower damage level than the south wall (Figure 2.11b). The rock mass was observed to be dry. Figure 2.12 showed the rock mass condition at Stop #2 after the rehab of the tunnel. Rock bolts were installed on the roof and on the left wall; also, the mesh support and shotcrete can be seen, which were used to prevent the falling of loose rocks. As previously mentioned, the backfilling activities took place on the right side of this location (Figure 2.2a). So the right wall of the tunnel shown in Figure 2.12 is the cemented rock fill.

Figure 2.12: Photo of the development drift and ground support at Stop #2
Source: From the mining company.

2.7 Summary

This chapter presents a general description of the mine site including geological settings and rock mass conditions, the in-situ stresses, the layout and geometry of tunnel system, the used ground control methods and the field measurements and observations.

It showed that the deposits are sediment-hosted and are generally controlled by the intersection of mineralized faults and stratigraphic units. The ore zones or altered rocks have low rock mass ratings, while the rock masses outside the ore zone are in better condition.

The tunnel system at the mine is complicated, driven in various heading directions and elevations. It also goes through the low rock mass quality area or the major fault zones. Primary ground control methodologies used at the mine include the rock and cable bolts, the shotcrete and mesh support, the cemented rock fill and so forth. Rehabilitation works have been conducted repeatedly by installing longer and stronger Swellex bolts. For the tunnels located in the very weak ground, the shotcrete arch support system was designed.

In aid of field instrumentations, major problems are figured out, which are likely to be at the wide tunnel intersections and the areas with low RMR ratings or areas of faulting. Because of the weak rock mass conditions, the tunnels normally exhibit a slow deterioration (i.e. yielding and unraveling).

Methodologies for investigating rock mass stability around underground excavations

3.1 Introduction

In this chapter, the major methodologies used for the investigation of rock mass stability around underground excavations are reviewed. The rock mass classification systems, which are empirical methods, are first introduced in the order they were developed with respect to time. For analytical methods, several available closed-form solutions for underground stability problems are provided. Following that is the description of the field instrumentations and the arisen numerical methods. The strengths and limitations of each method are then discussed and the final conclusions are made.

3.2 Rock mass classification systems

The rock mass classification systems are the empirical methods developed by researchers based on their experience from rock engineering projects. This approach is preferred to be used in the early stage of an underground project when limited information or analysis is available. With the development of several mature systems, they play important roles on the evaluation of rock mass conditions, the design of a preliminary support system, the implementation of the appropriate construction sequence and the determination of location or layout of openings (Stille and Palmström, 2003). The development of some remarkable rock mass classification systems is addressed hereunder.

The first successful attempt to classify the rock masses for engineering purposes was made by Terzaghi (1946). He proposed a simple rock load classification to estimate the needed support pressure from steel sets for tunnels. This classification system divided the rock masses into nine categories. The rock load factor (H_p) in each rock class, which is the height of the loosening zone above the tunnel roof, was given in relation with the tunnel width (B) and tunnel height (H_t). Terzaghi's classification gained wide applications since it was put forward. It was proved to be appropriate, although slightly conservative, for steel-supported rock tunnels (Cording and Deere, 1972). However, it is not suitable for modern supports, such as shotcrete and rock bolts (Bieniawski, 1989). Cecil (1970) stated that the classification described the rock condition too generally but with no quantitative information of the rock masses. According to Barton et al. (1974), Terzaghi's classification was useful for the medium-size tunnels; they also pointed out that the support pressure does not necessarily increase with increasing dimensions of the excavations. An important revision was made by Rose (1982) on the load coefficients for the fourth and sixth rock classes as little influence of the water table was considered.

Lauffer (1958) introduced the concept of stand-up time for an active unsupported tunnel span. The active unsupported span was defined as the width of the tunnel or the distance from the tunnel face to the support if it is less than the tunnel width (Bieniawski, 1989). The stand-up time is the length of time an unsupported tunnel can support itself before collapsing. It varies with the active span and the rock mass quality (Lauffer, 1958). Lauffer's original suggestion is no longer used, whereas the concept of stand-up time was used and modified in several later systems (Pacher et al., 1974; Barton et al., 1974; Bieniawski, 1993).

The Rock Quality Designation (RQD) developed by Deere et al. (1967) aimed to measure the rock quality based on core recovery by diamond drilling. The definition of RQD was the percentage of core recovered in intact pieces over 100 mm (4 in.) in length in the total length of the borehole, given by the following equation (Deere and Deere, 1988).

$$RQD = \frac{\sum \text{Length of core pieces} > 10\,\text{cm}\,(4\,\text{in.})}{\text{Total Core Run Length}} \tag{3.1}$$

Deere (1968) derived the relations between the RQD index and the engineering quality of the rock. Attempts were also made to correlate RQD with the tunnel support requirements by Deere et al. (1970), Cecil (1970), and Merritt (1972). Nevertheless, Merritt (1972) concluded that the support requirements estimated based on the RQD were not applicable to the problems where joints contain thin clay fillings or weathered material. As a simple index, RQD provides limited information about the rock quality; the influence of some critical features (i.e. the orientation and roughness of the discontinuities) cannot be taken into account by this index (Bieniawski, 1989). In addition, the RQD value changes with the direction and highly depends on the quality of drilling and site locations. It is therefore too risky to be used for engineering problems alone.

Nowadays, RQD is more popular to be used to estimate the rock mass properties (Zhang and Einstein, 2004; Zhang, 2016) and used as a basic parameter in the two widely used rock mass classification systems: the rock mass rating (RMR) (Bieniawski, 1976, 1989) and Q system (Barton et al., 1974). Under the circumstances when the RQD cannot be obtained directly from the core, some equations were put forward associated with the fracture frequency. For example, Palmström (1982) suggested estimation of the RQD from the number of joints per unit volume for clay-free rock mass, given by

$$RQD = 115 - 3.3J_v \tag{3.2}$$

Where J_v is the total number of joints per cubic meter.

The Rock Structure Rating (RSR) system was developed by Wickham et al. (1972) to predict the support requirements for tunnels using the geologic and construction information. The system gives a numeric scale of 0 to 100, which is the sum of weighted ratings determined by three parameters (Wickham et al., 1972). The first parameter is related to the rock type and geologic structure; the second parameter involves the joint spacing and joint orientation with respect to the direction of tunnel drive; the third parameter correlates the joint condition and groundwater. The RSR was the first quantitative and complete rock mass classification system; the major contribution was the usage of ratings to represent the importance of the input parameters (Bieniawski, 1989). However, since most of the case histories that the system based on were the steel-arched tunnels, the classification seems more appropriate for the design of steel supports in tunnels (Bieniawski, 1989).

Bieniawski (1973) developed a quantitative classification for the rock masses called Geomechanics Classification or the Rock Mass Rating (RMR) system. It is one of the commonly used classifications nowadays. The classification of the rock mass was based on the ratings of six parameters (Bieniawski, 1988), including the uniaxial compressive strength of rock material, the rock quality designation (RQD), the spacing of discontinuities, the condition of discontinuities, the groundwater conditions and the orientation of discontinuities. The final summed value (RMR value) divides the rock masses into five classes and can be used to estimate the stand-up time, the maximum stable rock span and the support requirements (Bieniawski, 1989). It had been modified over the years by adding new data from case histories and been extended to multiple fields such as tunneling, mining, dam foundations and slope stability (Bieniawski, 1989; Gonzalez de Vallejo, 1983; Laubscher, 1977, 1984; Serafim and Pereira, 1983; Romana, 1985). The RMR value has also been used for estimating the rock mass properties. For instance, Bieniawski (1989) and Serafim and Pereira (1983) proposed the relations between the rock mass deformation modulus and the RMR values. The RMR was utilized in the determination of Hoek-Brown constants (Hoek and Brown, 1980, 1988).

Another widely applied classification is the Q system introduced by Barton et al. (1974) in Norway. The system was based on the data of 212 tunnel cases (Barton et al., 1974). The Q value was determined by six parameters, as given by the following equation.

$$Q = \frac{RQD}{J_n} \cdot \frac{J_r}{J_a} \cdot \frac{J_w}{SRF} \tag{3.3}$$

where RQD is the rock quality designation, J_n is the joint set number, J_r is the joint roughness number, J_a is the joint alteration number, J_w is the joint water reduction factor, and SRF is the stress reduction factor.

According to Barton et al. (1974), the three quotients in Eq. (3.3) represent the measures of the relative block size $\left(\frac{RQD}{J_n}\right)$, the inter-block shear strength $\left(\frac{J_r}{J_a}\right)$, and the active stress $\left(\frac{J_w}{SRF}\right)$. The parameters J_r and J_a were taken from the unfavorable joints, and the system was intended to deal with general cases rather than the problems involving special conditions such as clay-bearing weakness and fault zones (Barton et al., 1974). The support requirements were given by the chart in relation to the equivalent dimensions of the excavation and the Q value (Barton et al., 1974). The support chart was updated by Grimstad and Barton (1993) by adding new case histories with the usage of steel-fiber reinforced sprayed concrete.

The Geological Strength Index (GSI), developed by Hoek (1994) and Hoek et al. (1995), provided an indicator to account for the reduction in intact rock strength due to different geological conditions. This characterization is simply based on the visual impression of the rock structure and the surface condition of the discontinuities (Hoek and Brown, 1997). Once the GSI value is estimated by the field observation, it can be used to calculate the rock mass strength parameters and the rock mass deformation modulus (Marinos and Hoek, 2000, 2001; Cai and Kaiser, 2004; Cai et al., 2007a; Sonmez et al., 2004). Hence, instead of being considered as a rock mass classification, the GSI system is more appropriate to be used as an index to estimate the rock mass parameters.

In summary, based on the field observations, the laboratory or in-situ tests, the rock mass classification systems are easy to apply and suitable to be used at an early stage of construction. Wide applications of the rock mass classification systems in rock engineering projects showed that they are useful in the evaluation of rock mass conditions, the design of preliminary supports as well as the estimation of rock mass properties. Nevertheless, they should never be attempted to guide the ultimate design or to replace other sophisticated methods (Bieniawski, 1988). The stability assessment of rock masses, the knowledge of intrinsic mechanisms and possible failure modes are out of the capability of the classification systems. Although several input parameters are used in the recent rock mass classification systems such as RMR and Q, some key factors are missing, such as the explicit representation of discontinuity, orientation and size, the effects of blasting, staged tunneling processes and time-dependent deformation. On the other hand, even though the same output rating is obtained, the rock mass behavior could be diverse due to various combinations of classification parameters (Stille and Palmström, 2003). The output description or a single number from the classifications is inadequate for support design and stability evaluation in complex geological conditions (Riedmüller and Schubert, 1999; Stille and Palmström, 2003). For a complicated underground engineering problem, at least two systems should be applied for the preliminary design, while conjunction with other approaches such as analytical methods, field observations and numerical methods are needed for further assessment (Stille and Palmström, 2003).

3.3   Analytical methods

In analytical methods, the closed-form solutions are normally derived based on theories or laws. It is probably the oldest and most accessible approach. In underground rock engineering, many solutions have been developed to assess the rock mass stabilities and to predict the deformations and failures around the excavations.

For instance, some theories are available for the openings excavated in the laminated rock masses in mining engineering. It was proposed by Fayol (1885) that the stratified roof strata could be assumed as a stack of simply supported beams; the weight of each beam layer was transferred to the abutments of the opening rather than being carried by the beams underneath, generating an arch structure above the mine opening. The beams were simplified as continuous with no joints and hence were proved to rupture in the mid-span (Bucky, 1934; Bucky and Taborelli, 1938). The closed-form solutions based on this theory can be found in the studies of Please et al. (2013), Jiang et al. (2016), Zhang et al. (2016) and others. As the crosscut joints were considered, the voussoir beam theory (Evans, 1941) was proposed and used to seek the solutions. Evans (1941) established the relation between the vertical deflection and the lateral trust of a jointed voussoir beam. Sofianos (1996) derived the formula to calculate the deflections and strains of the hard rock roof based on his voussoir beam model; the relations associated with the loading, the mechanical parameters and the dimensions of the beam were obtained for the three failure modes. Diederichs and Kaiser (1999) made improvements on the classic voussoir analogue model and came up with a yield limit; the hanging walls at two mines were analyzed and the predicted displacements were verified using the field monitoring data.

For simple excavations (i.e. spherical and cylindrical (circular) tunnels), analytical solutions have been developed to calculate the stresses, strains, displacements and yielding zones. They were derived based on the intrinsic law of the material (i.e. the constitutive relations and/or the assumed failure criterion. For instance, Kirsch (1898) gave the elastic solution (Kirsch's solution) of the tangential and radial stresses and displacements around a circular opening, which is still widely used today. Analytical studies for rock masses with plastic behaviors, such as elastic-brittle-plastic, elastic-perfectly-plastic, and strain-softening, are also available (Brown et al., 1983; Carranza-Torres and Fairhurst, 1999; Sharan, 2003, 2005; Zhang et al., 2012). Additionally, some complex solutions considering the influences of the time-dependent behavior, in-situ stresses, water pressure, and rock supports can be found in the studies of Goodman (1989), Lu et al. (2010), Fahimifar and Reza Zareifard (2009), Bobet and Einstein (2011) and others.

The convergence-confinement method is a theoretical tool to design the support for underground excavations (Alejano et al., 2010). It includes three essential curves, the ground response curve (GRC), the longitudinal deformation profile (LDP) and the support characteristic curve (SCC). By constructing the three curves, the response of the rock masses and the interactions between the rock mass and supports can be predicted. Carranza-Torres and Fairhurst (2000) applied the convergence-confinement method for a tunnel in the rock masses that satisfy the Hoek-Brown failure criterion; the SCC curves were obtained for the practical supports including the shotcrete or concrete rings, the blocked steel sets, and the ungrouted bolts and cables. Alejano et al. (2010) put forth the procedure to construct the GRC curves for the strain softening rock masses; the proposed method was used to predict the required supports for the tunnels in various rock mass conditions.

The analytical methods have been applied for the analysis of simple underground problems; the relevant failure mechanisms or behaviors of rock masses can be studied to a certain degree. Nevertheless, the solved problems are generally for cases with homogenous, isotropic rock material, axisymmetric excavation, plane-strain assumption and so forth. It is difficult or impossible to derive the closed-form solutions for complex problems; instead, many equations have been solved by the aid of computer programs (Guan et al., 2007; Fahimifar and Soroush, 2005).

3.4   Field instrumentation methods

Field instrumentation is normally applied at all stages of an underground project for different kinds of purposes. Prior to the construction, it is used to collect the required information for design, such as the types and properties of the rock, the conditions and orientations of the structures, the direction and magnitude of in-situ stress and so forth. During or after the excavation, field instrumentation plays an important role on monitoring the rock mass response, predicting the potential failures and modifying the construction and supporting procedures (Hoek and Brown, 1980). With respect to assess the stability of underground structures, it usually involves the measurement of displacements, stresses, strains and pressures which can directly reveal the rock mass behaviors, or of the indirect parameters, such as micro-seismic activities, ultrasonic velocities, permeabilities and so forth.

The reveal of rock mass behavior and stability analysis of underground excavations using field measurements have been implemented in numerous projects. For instance, Kimura et al. (1987) studied the settlement behavior and possible shear failure for a shallow tunnel by monitoring the shear strain and maximum surface settlement. Kavvadas (2005) introduced the different safety goals and the various instruments of ground deformation measurement for shallow (urban tunnels) and deep mountain tunnels. In addition, the deformation monitoring results of three case studies were analyzed in detail; their usage in evaluating the safety of ground surface structures, in designing the support system and in predicting the potential collapse were illustrated. Bruneau et al. (2003) investigated the influence of faults on the stability of a mine shaft by the measurements using various instrumentations, such as glass slide, feeler gauge, LVDTs, fishplate and potentiometer. The shaft degradation was proved by the vertical displacements across the faults (cracking in the walls) and by the deformations of the shaft steel guide rails and concrete lining at different levels. Satyanarayana et al. (2015) recorded the strata load, excavation convergence, induced vertical stress and roof deformations to study the strata behavior during depillaring in an underground coal mine.

The Underground Research Laboratory (URL) was constructed by Atomic Energy of Canada Limited (AECL) to study the issues related to the deep geologic disposal of used fuel from nuclear reactors (Martino and Chandler, 2004). A large amount of field monitoring research has been performed at URL to investigate the excavation response of the rock masses (Read, 2004). The mechanical behavior of the rock masses around the excavations was evaluated through the monitoring of the axial displacements in boreholes, the excavation convergence, the stress changes, the three-dimensional displacements and the time-dependent deformations (Martin and Simmons, 1993). The coupled hydro-mechanical behavior was investigated using the data of in-situ stresses, the permeability and normal stiffness of the fractures (Martin and Simmons, 1993). The Mine-by Experiment (MBE) at URL aimed to study the excavation-induced damaged development and progressive failure around an underground opening subjected to high differential stresses under ambient temperature conditions (Read, 2004). The details of some related studies have been performed by Read and Martin (1992), Falls and Young (1998), Maxwell et al. (1998) and Martino and Chandler (2004). One important objective of these studies was to estimate the excavation damaged zone (EDZ), which contains induced microfractures and can provide the enhanced permeability pathways for radionuclide migration. By monitoring the acoustic emission (AE) activities and the changes of ultrasonic velocity, the density of micro-cracks and the extent of EDZ could be obtained.

Displacement has been considered as the most effective indicator of the rock mass behavior for underground excavations (Hoek and Brown, 1980). The measurements commonly involve the convergence by rod or tape extensometers and the displacement in the surrounding rock mass by borehole extensometers. To serve as the alarm to predict the possible failure during and after construction, long-term monitoring is needed. However, the field condition always turns out difficult in providing continuous and reliable measurements. Under such conditions, the back analysis based on limited data is an alternate solution. By using the field measurements (i.e. displacement), which should be the output of the forward analysis, as the input, in-situ properties of the rock mass or in-situ stress field can be estimated and subsequently used in the further design or analysis (Sakurai, 1997). Sakurai and Takeuchi (1983) back calculated the initial stress and material constants according to the measured displacements, and these parameters were used as the input data in the finite element analysis to determine the strain distribution around the tunnels. Based on the field monitoring data at URL as mentioned above, many back analyses were carried out to determine in-situ modulus (Lang et al., 1987) and the state of in-situ stress (Kaiser et al., 1990; Thompson and Chandler, 2004). Similar studies were also given by the researchers, such as Kristen (1976), Feng and Lewis (1987) and Yazdani et al. (2012). In addition, the field instrumentation can be used as a feedback to evaluate the design of the support and construction method, or to verify the accuracy of existing analysis (Sakurai, 1997; Kovari and Amstad, 1979).

By applying the field instrumentations, various information on the rock mass behavior can be obtained. However, due to the uncertainties and complexities that appeared in the field, continuous and reliable monitoring data is sometimes unavailable, which may result in the ineffective assessment of rock mass stability. The prediction of rock mass response could be compromised without the knowledge of failure mechanisms. On the other hand, field monitoring data could be utilized in the back analysis to estimate in-situ parameters or for the verification of the existing analysis.

3.5   Numerical methods

With the advent of advanced computer techniques, the numerical method has become an efficient and low-cost tool to solve the complex problems in rock engineering. For a practical problem, the rock mass response to construction disturbance is complicated, affected by various factors. Numerical modeling can capture the most relevant mechanisms of the system and provide adequate models for various engineering purposes (Cundall, 1971, 1980, 1988; Hart et al., 1988; Shi, 1988; Itasca, 1993; Jing, 2003). The commonly used numerical methods in rock engineering are categorized as continuum, discontinuum and hybrid approaches.

In continuum methods, the rock material is considered as continuous; the discontinuities can be modeled as the elements with different material properties from the intact rock or as the special joint elements. Three well-known continuum methods are the Finite Difference Method (FDM), the Finite Element Method (FEM), and the Boundary Element Method (BEM).

The FDM appeared as the earliest among the three methods. Its basic concept is to solve the partial differential equations by replacing the partial derivatives with differences defined at neighboring grid points (Itasca, 1993). The solution of the equations is direct and efficient without forming global matrix equations. This distinct feature makes FDM good at dealing with non-linear behavior of the rock materials. The representative computer code based on FDM is the Fast Lagrangian Analysis of Continua (FLAC), developed by Itasca (Itasca, 1993). Applications of FLAC code for underground design or stability analyses can be found as follows. Alejano et al. (1999) used FLAC to estimate the subsidence due to the flat and inclined coal seam exploitation; in their research, the material behaviors including transversely isotropic elastic pre-failure behavior, anisotropic yield surface, and the isotropic elastic post-failure behavior were simulated. Hsiao et al. (2009) investigated the influence of different factors including rock strength, rock mass rating, overburden depth, and intersection angle on the roof settlement of a tunnel intersection using FLAC3D; the stability of the tunnel was evaluated based on an empirical safety criterion. According to the simulation results, the supporting strategies were proposed with respect to different geological conditions, different intersection angles and different locations in the intersection area. Mortazavi et al. (2009) studied the non-linear behavior of rock pillars in underground openings using FLAC; stresses and deformations of the pillars were computed for various cases with varying pillar geometry and pillar residual strength. Xu et al. (2013) developed a three-dimensional model including different strata and fault zones to predict the surface subsidence of a coal mine in China. The FLAC code plays an important role on solving engineering problems; it is appropriate for simulating the non-linear behaviors (plasticity, strain softening, etc.) of the rock masses rather than that of the discontinuities.

Unlike FDM, the FEM uses the implicit solution scheme. After the discretization, the local assumptions are made by selecting appropriate shape functions (interpolation functions) of the unknown variables to satisfy the element governing equations. The global equations are formed and need to solve. FEM is widely used for the rock mechanics problems in civil engineering due to the ability to handle material heterogeneity, non-linear deformability and complex boundary conditions, in-situ stresses and gravity (Zienkiewicz et al., 1970; Ghaboussi et al., 1973; Kulatilake, 1985; Gens et al., 1995; Yoshida and Horii, 1998; Jing and Hudson, 2002). The concept of representation of rock fractures was introduced since the late 1960s. Goodman et al. (1968) pointed out the inaccuracy of joint characterization by the plane-strain continuum element; instead, they proposed the zero-thickness joint element and derived the joint element equations, and subsequently, the joint model was applied to simulate the sliding of a joint with a saw-tooth, the block movements and rotations at the joint intersections, and the roof collapse of a blocky tunnel. After that, different joint element models were developed by some other researchers (Zienkiewicz et al., 1970; Ghaboussi et al., 1973; Gens et al., 1995). Nevertheless, quite a number of applications of FEM incorporated joint effects implicitly to avoid the computational complexities. For instance, in the study of Yoshida and Horii (1998), the joints were considered as microstructure and the jointed rock masses are modeled using the equivalent continuum model; the proposed method was applied to the excavation of large-scale caverns. Sitharam et al. (2001) performed the equivalent continuum analysis by incorporating the effect of joints using a joint factor, which involves the joint spacing, joint orientation and joint condition; then the displacements around a large underground opening were predicted based on the equivalent continuum model. Lee and Song (2003) used the rheological model which includes the units of joint sets and rock bolts to investigate the stability of an underground cavern. With the development of powerful hardware and software, many computer programs, based on the FEM, are now available for solving rock engineering or rock mechanics problems. The RFPA introduced by Tang (1995) was used to model the progressive failure and associated seismicity in brittle rock or rock mass. Phase^2 (Rocscience Inc., 2004) is the numerical tool designed for the finite element analysis and support design for excavations. The program ABAQUS can take care of the non-linear behavior of the material (Huang et al., 2013) as well as the dynamic failure (Sazid and Singh, 2013). PLAXIS is the finite element code for the analysis of deformation and stability for geotechnical engineering projects (Cui et al., 2007).

BEM is a general numerical technique which solves boundary integral equations. The solution procedures include the discretization of the boundary, the approximation of the solution of functions locally at boundary elements by shape functions, the incorporation of boundary conditions and the estimation of displacements and stresses inside the domain (Brady and Bray, 1978; Crouch and Starfield, 1983; Beer and Poulsen, 1994; Pan et al., 1998; Chen et al., 1998). The technique of solving the integral equations makes the BEM more accurate than the FDM and FEM (Blandford et al., 1981; Jing, 2003). Due to the fact that BEM only discretizes the boundary, it has the advantage in handling the problems of crack propagation with large or infinite domains. In the research of Chen et al. (1998), BEM was used to simulate the diametral compression tests on the initially cracked discs; the fracture toughness, the angle of crack initiation and the path of crack propagation were predicted. Pan et al. (1998) presented the applications of BEM for solving the anisotropic half-plane problems; the influence of material anisotropy on stress distributions around the tunnels and in the ground was given. Beer and Poulsen (1994) developed a method with BEM to simulate rock joints and faults, where the problem domain was divided into several boundary element regions and the joint behavior was assigned for the interfaces between regions; two simple applications, the mine excavation with a fault and the geological modeling of lithological contact and faults, were investigated. Using the boundary element program MAP3D, a mine model involving complex geometries was modeled by Kaiser et al. (2001) to predict the stress changes during stope excavation. Chu et al. (2007) investigated the mechanical behavior of a twin-tunnel in the multi-layered formations; the layered formations were realized by using the boundary element program FSM.

Continuum approaches, as discussed above, have been successfully applied in rock mechanics and rock engineering for many decades. Nevertheless, the continuous assumption means that the material will never be open or broken into pieces; the joint displacements are restricted to small values (Cundall, 2001; Itasca, 2007; Jing, 2003). Hence, they are more appropriate for the problems where the rock is relatively intact or where the rock mass is highly fractured (equivalent continuum situation). In fact, most rock failures of underground constructions are related to natural defects. To capture the realistic behavior of discontinuities, a method that can account for large displacements or rotations is necessary.

The discontinuum modeling, represented by the distinct element method (Cundall, 1971, 1988) and the discontinuous deformation analysis (DDA) (Shi, 1988), considers the medium as an assemblage of discrete blocks which are connected by contacts or interfaces. The discontinuities are explicitly represented and their behaviors can be described with specific joint constitutive models. This method allows for the large displacements and rotations of the discrete blocks, including the complete detachment. Several representative numerical codes and their applications are summarized as follows.

In the distinct element method, the rock mass can be modeled as rigid or deformable blocks; the discontinuities are explicitly modeled as distinct boundary interactions. This method utilizes the explicit solution scheme so that no matrix needs to be formed during the calculation process. Thus, it is termed as the explicit approach of the discrete element method (DEM) (Cundall, 1980, 1988; Itasca, 2007). Because of the adoption of an explicit scheme, the distinct element method is capable to accommodate complex constitutive behavior for both the intact material and discontinuities (Itasca, 2007). The two distinct element codes, UDEC (Cundall, 1980) and 3DEC (Cundall, 1988; Hart et al., 1988), are used for 2-D and 3-D numerical modeling, respectively. Their applications on underground excavations have been addressed by numerous researchers. For instance, the UDEC was used as the tool by Chryssanthakis et al. (1997) to investigate the effects of the fiber-reinforced shotcrete and the construction sequences on the stability of the tunnels. Bhasin and Hoeg (1997) studied the rock mass behavior of a large cavern by varying the joint spacing and joint strength parameters. Shen and Barton (1997) used UDEC to explore the influences of joint spacing and joint orientation on the disturbed zones (failure zone, open zone and shear zone) around the tunnels. Similarly, Hao and Azzam (2005) used UDEC to investigate the effects of fault parameters on the rock mass behavior around underground excavations. Gao et al. (2014) simulated the roof shear failure in coal mine roadways by developing a FISH function in UDEC to track the newly formed contacts during excavation; also the effect of rock bolts on the roof behavior was studied. Wang et al. (2012) performed three-dimensional analyses using discontinuum and continuum models in 3DEC; multiple studies were performed regarding the effects of the discontinuities, the geomechanical parameters of rock masses and discontinuities, and the rock support system. Similarly, Wu and Kulatilake (2012b) conducted the 3-D stress analyses on the tunnel stability using an equivalent continuum/discontinuum model and a fully discontinuum model. Using the joint data collected by laser scanning (Lidar), Fekete and Diederichs (2013) developed a discontinuum model using 3DEC to simulate the structurally controlled failure around a tunnel in a blocky rock mass. In the research of Shreedharan and Kulatilake (2016), 3DEC was used to investigate the stability of the tunnels with two different shapes in a deep coal mine in China; the effectiveness of the support system was evaluated by implementing the instantaneous and delayed installation of rock supports. Sherizadeh and Kulatilake (2016) evaluated the effect of various factors including the variation of rock mass strength properties, the variation of discontinuity mechanical properties, the orientations and magnitudes of the horizontal in-situ stresses, and the size of pillars and excavations on the roof stability in a coal mine by using 3DEC. By developing a 3-D discontinuum model, Huang et al. (2017) numerically simulated the ground subsidence due to the extraction and backfilling during open stopping operations. Cui et al. (2016) carried out a seismic analysis for an underground chamber by using 3DEC. In their numerical model, a large geological discontinuity was modeled and assigned with the continuously yielding joint model; they found that the discontinuity had a major influence on the deformations around the chamber.

DDA was originally developed to back-calculate Young's modulus and the Poisson's ratio from displacement measurements (Shi and Goodman, 1985), and later extended to perform large displacement analysis for blocky rock masses (Shi, 1988). Different from the distinct element method, DDA is based on the implicit solution scheme (the implicit approach of the DEM), similar to that used in the FEM. The displacements are the unknowns to be solved for. The interaction between the blocks is simulated by mechanical springs or penalty functions and the system of simultaneous equations is obtained by minimizing the total energy of the whole system (Khan, 2010). The implicit algorithm has its superiority that the calculations are unconditionally stable with the time-step size limited only by accuracy considerations. It is also one of the advantages of DDA over the distinct element method suggested by some researchers (Jing, 2003; Khan, 2010). However, the larger time step does not necessarily mean the higher convergence speed (Khan, 2010). It was also shown that the appropriate parameters in DDA are difficult to be determined for an arbitrary problem and the accuracy cannot be guaranteed under some circumstances (Khan, 2010; Ohnishi et al., 2014; Shreedharan and Kulatilake, 2016). The applications of DDA covered multiple topics, including tunneling, caverns, fracturing and fragmentation processes of geological and structural materials and earthquake effects (Jing, 2003). Many studies of DDA have focused on the two-dimensional modeling (Yeung and Leong, 1997; Kim et al., 1999; Wu et al., 2005; Tsesarsky and Hatzor, 2006; Shi, 2014; Chen et al., 2016). Due to the rigorous scheme that DDA uses, the difficulty for 3-D DDA is the development of a complete contact theory that governs the interaction of many 3-D blocks (Yeung et al., 2007). So far, a lot of studies have been performed on the establishment of an efficient contact detection strategy (Jiang and Yeung, 2004; Wu et al., 2005; Yeung et al., 2007; Beyabanaki et al., 2008; Ahn and Song, 2011). Even though the three-dimensional modeling has been successfully implemented in recent years for underground openings (Yeung et al., 2008; Shi, 2014; Zhu et al., 2016), most of the problems being solved were simple cases; the complex behavior of discontinuities is difficult to be simulated.

The hybrid method was developed to take advantage of the strength of different methods and to avoid the disadvantages of the multiple methods. The commonly used hybrid models in rock engineering include the BEM/FEM, DEM/BEM and DEM/FEM models. As previously mentioned, the BEM involves discretization of the boundaries and may largely reduce the calculation time. It usually undertakes the role of modeling far-field medium, while the FEM or DEM is used to simulate the non-linear, anisotropic or discontinuous domains. For example, the hybrid BEM/FEM was applied by Swoboda et al. (1987) and Eberhardsteiner et al. (1993) for the two- and three-dimensional analyses of tunnel excavations; the near-field places where the stress concentrations and plastic deformations are likely to occur, such as interior of the tunnel, the shotcrete shell and its outer vicinity, were discretized by the FEM; the far-field rock masses, on the other hand, were simulated as elastic by the BEM. In a similar way, the DEM/BEM method aims at the problems where the explicit representation of discontinuities is needed in the interested region, surrounded by the elastic far-field rock masses (Lorig et al., 1986; Feng et al., 1998; Chen and Zhao, 2002). The finite-discrete element method has been developed by some researchers to capture the fracturing process of rock masses (Munjiza et al., 1995; Munjiza et al., 1999; Klerck, 2000; Klerck et al., 2004). Instead of simply discretizing problem domains with different elements, this method involves a different couple mechanism, where the system is comprised of a number of separate deformable bodies (discrete elements) and the discrete elements are discretized into finite elements (Munjiza et al., 1999). The fracturing of continuum media (transition from continuum to discontinuum) and the interaction of the generated fragments can be modeled (Mahabadi et al., 2010). Some finite-discrete element codes, such as Y-Code (Munjiza et al., 1999), Y-Geo (Mahabadi et al., 2012) and ELFEN (Rockfield Software Ltd, 2011), have been developed. The major applications fall in rock mechanics, such as modeling laboratory tests (Cai and Kaiser, 2004; Karami and Stead, 2008). The engineering applications in engineering problems, such as rock slope and underground stability, have been implemented for simple cases (Eberhardt et al., 2004; Barla et al., 2011, Sellers and Klerck, 2000; Vyazmensky et al., 2007). The major difficulties still exist in the computational demand for solving complex engineering problems (Mahabadi et al., 2012) and some unavailable aspects of rock mass modeling (Elmo, 2006).

In conclusion, each numerical method is appropriate for specific rock problems while it is limited to some others. To solve a problem properly using numerical modeling, the suitable means need to be determined by aware of its strengths and weaknesses. Coggan et al. (2012) summarized the capabilities and limitations of most of the aforementioned numerical approaches, as given in Table 3.1. Generally, the continuum method, which follows the continuous assumption, is suitable for the excavations in intact rock or highly fractured rock masses. The latter system where the discontinuities are ubiquitously distributed is normally solved using the equivalent continuum modeling. The discontinuum method, on the other hand, is preferred to handle problems in moderately jointed rock masses or where the behavior of the discontinuity is indispensable (Cundall, 2001). The hybrid method, which seems to be promising, is still under development with challenges in adding new features and seeking proper algorithms. A more comprehensive description of the methods was given by Jing (2003) and Coggan et al. (2012). The selection of an appropriate tool depends on the problem-specific factors, including the problem scale, fracture geometry system and so forth (Jing, 2003; Brady and Brown, 2004). Certainly, the experience and knowledge of the researcher are also important, determining what features are essential and should be incorporated with respect to the specific problem.

Table 3.1 Capabilities and limitations of the numerical methods for the analysis of underground excavations[a]

Type of numerical method	Capabilities	Limitations
BEM	The capability of three-dimensional modeling Rapid assessment of designs and stress concentrations	Normally elastic analysis only (non-linear and time-dependent options are available)
FEM and FDM	Allow for material deformation and failure, can model complex behavior; The capability of three-dimensional modeling able to assess and simulate both saturated and unsaturated (multiphase) flow/water pressures Recent advances in hardware mean that complicated models can now be PC-based and run in reasonable time periods Can incorporate coupled dynamic/groundwater analysis, time-dependent deformation readily simulated	Must be aware of model/software limitations including effects of mesh size, boundaries, symmetry and hardware restrictions (i.e. memory and time constraints) and data input limitations (such as effects of variation of critical input parameters, etc.) Deformations along discontinuities are restricted to small values; well-trained and experienced users and familiarity with numerical analysis methods essential Validation through surface/subsurface instrumentation important
DEM	Able to model complex behaviors including both block deformation and relative movement of blocks (translation/rotation) Three-dimensional models possible Effect of parameter variations on instability can be investigated easily Dynamic loading, creep and groundwater simulated Can incorporate synthetic rock masses to represent the fracture network Use of Voronoi polygonal blocks allows simulation of rock fracture between blocks	Must be aware of model/software limitations including effects of mesh size, boundaries, symmetry and hardware restrictions (i.e. memory and time constraints) and data input limitations (such as effects of variation of critical input parameters, etc.) Scale-effects: simulate representative discontinuity geometry (spacing, persistence) Limited data on joint stiffness available Validation through surface/subsurface instrumentation important
Hybrid codes (finite-discrete element)	Able to allow for the extension of existing fractures and creation of new fractures through intact rock Capable of three-dimensional modeling (although limited application to-date) Can incorporate dynamic effects	Limited use and validation State-of-the-art codes requiring in-depth knowledge/experience of modeling methods/mechanics Must incorporate realistic rock fracture network Little data available for contact properties and fracture mechanics properties Limited capability to simulate the effects of groundwater Extremely long run times will require the use of parallel processing for large models

[a] Slightly modified from J. Coggan, F. Gao, D. Stead & D. Elmo, "Numerical modelling of the effects of weak immediate roof lithology on coal mine roadway stability," *International Journal of Coal Geology*, 90 (2012): 100–109.

3.6 Summary

In projects involving rock construction, the rock mass classification systems form the backbone of the empirical design methods. By considering the combined effect of different geological parameters, the rock mass classifications are effective to evaluate the overall rock mass quality, to provide preliminary support guidelines for underground excavations and to estimate the rock mass properties. Since they were developed based on the experience from previous projects and due to the fact that some critical factors are disregarded, the use of rock mass classification systems should be limited to simple situations. Accurate rock mass behavior or intrinsic rock failure mechanisms should be explored by using more rigorous methods.

The analytical methods, which provide theoretical solutions, may be applicable for underground problems having very simple geologic media. The rock mass behavior and failure mechanisms may be captured by this method to a certain degree. It seems, however, difficult to be used in solving problems with complex geologies and engineering disturbances.

The field instrumentation is essential for underground engineering projects since it provides valuable information on the rock mass response, either directly or indirectly. It is a useful aid to predict the high rock mass movements and potential rock failures. Yet due to the unfavorable geological or in-situ conditions, the continuity and reliability of field monitoring data cannot be guaranteed. The obtained results are also site-dependent. Alternative uses of the field measurements can be provided for the back analysis or the validation of existing analysis.

Numerical modeling has been widely used in the design or stability assessment of rock engineering structures. This method is powerful to take care of the complexities, including the uncertainties of rock mass and discontinuity properties, the various geometry patterns of discontinuities, the complex rock mass and discontinuity constitutive behaviors, the construction and supporting effects and so forth. Depending on the specific problem that needs to be solved, a suitable numerical tool has to be selected. A good understanding of the strengths and limitations of the numerical codes and of the specific site conditions is required.

Consequently, the aforementioned methods play different roles during the process of construction of rock engineering structures. Due to the complex geological conditions and mine constructions, the variability and uncertainty in estimating mechanical properties and so forth, the assessment of rock mass stability for an underground mine is really challenging and difficult. The numerical modeling, which can capture the most relevant mechanisms, seems to be capable of tackling this difficult problem. It should be applied wisely and in conjunction with other methods (i.e. the empirical methods and the field instrumentations) for the purpose of developing an adequate model.

Chapter 4

Some critical factors in modeling of rock mass stability around underground excavations

4.1 Introduction

According to the previous chapter, numerical modeling turns out to be a capable method for stability assessment of underground rock structures. The application of this method in rock engineering is both a science and art for the purpose of developing an adequate model. Due to the fact that various complexities may exist with respect to the geological conditions and the engineering constructions, the most relevant and important factors need to be figured out. In this respect, the following sections cover the reviews and discussions on the essential aspects that need to be captured by numerical modeling of rock mass stability around underground excavations.

4.2 Representation of the discontinuities

As a natural geological material, the rock masses contain different types of discontinuities, in terms of small-scale ones such as fissures, fractures and joints, or of large-scale faults such as bedding planes, shear zones and dikes. These pre-existing defects make the rock mass discontinuous, anisotropic, inhomogeneous, and non-linear, and highly weakened with respect to the deformability and strength of the material (Kulatilake et al., 1985, 1993b; Aydan et al., 1997; Jing, 2003). When an underground opening is made, these discontinuities play a major role with respect to the instabilities of surrounding rock masses. Hence, the knowledge of mechanical behavior of the discontinuities and of their influence on the rock mass stability is necessary for the analysis of underground stability problems.

As introduced in Chapter 3, the DEM is the numerical method for the explicit representation of discontinuities. Two crucial issues related to joint representation are respectively the joint geometry data and joint material properties (Kulatilake et al., 1985, 1993a, 1993b; Barla and Barla, 2000). To incorporate realistic joints in the field, certain efforts have been made by previous researchers. Kulatilake (1998) developed the software package FRACNTWK to analyze the discontinuity data obtained from boreholes, rock cores, scanlines and 2-D exposures, such as rock outcrops, tunnel walls and tunnel roofs; by using this software, the number of joint sets and the density, orientation, size and the location distribution of the joints can be estimated. Even prior to 1998, computer programs from this software package were used to model fracture systems in 3-D, including verifications for a number of real-world projects (Kulatilake et al., 1993a, 1996). Kulatilake et al. (2003) illustrated the detailed procedures to characterize the joint sets based on the scanline surveys and borehole fracture data for Arrowhead East Tunnel in California; a 3-D fracture system was formed and validated using the field mapping data. Using the same software package,

Wu and Kulatilake (2012a) estimated the 3-D statistics of fracture size, orientation and intensity for a dam site in China and performed discontinuum tunnel numerical modeling incorporating these fracture sets. Elmo and Stead (2010) and Benedetto et al. (2014) performed some discrete fracture network modeling using some commercially available fracture network software. Other efforts have been made by Fekete and Diederichs (2013) to reconstruct the fracture network using the deterministic and statistical interpretations of the collected Lidar data. They pointed out that Lidar scanning provides data points rather than continuous data so that the joint persistence is difficult to estimate and stressed the importance of discontinuum rock mass models for understanding the failure mechanisms and sensitivities of the practical problem. With respect to the joint persistence, it has been recognized as a critical factor affecting the deformation, strength and failure modes of the jointed rock masses (Brown, 1970; Kulatilake et al., 1993b, 1997; Prudencio and Van Sint Jan, 2007). Simulation of realistic geometry of joints is essential for the stability analysis of rock structures. Kulatilake et al. (1992) proposed a procedure to incorporate finite-sized joints into a discrete numerical model by combining them with fictitious joints to divide the problem domain into polygons in two dimensions or into a polyhedral in three dimensions to perform UDEC modeling in 2-D or 3DEC modeling in 3-D.

Accurate estimation of joint mechanical properties is another critical and challenging task. The mechanical properties of joints can be influenced by a variety of factors, such as the joint contact area, the joint aperture, the roughness of the joint wall and the relevant properties of the filling material (Goodman et al., 1968). Reliable estimation of joint properties may be obtained from large-scale in-situ tests, but the cost is high and the selection of the appropriate testing size remains questionable to a given problem (Barton, 1972). On the other hand, a large number of studies have been carried out in the laboratory to investigate the mechanical behavior of joints. The deformability and strength behaviors of joints depend on the normal stress (Goodman, 1974; Barton and Choubey, 1977; Bandis et al., 1983); the magnitude of the deformability parameters, that is the joint normal and shear stiffness (Goodman et al., 1968) could vary in a range of several orders as a function of the normal stress (Barton and Choubey, 1977). Various models have been suggested to fit the variations of the normal or shear stiffness with the normal stress (Bandis et al., 1983; Swan, 1983; Malama and Kulatilake, 2003; Kulatilake et al., 2016). Studies by Bandis et al. (1981) and Barton et al. (1985) showed significant scale effects on the shear strength and deformability properties of rough and undulating joints. As the joint length increases, the peak shear displacement gets larger; the shear stress–normal stress relation changes from "brittle" with the high shear strength to "plastic" with lower strength; it was concluded that the ultimate strength of the laboratory-size joints is normally higher than that of the joints exposed in the field (Bandis et al., 1981).

Due to the aforementioned difficulties and uncertainties, extensive numerical studies have been performed to investigate the influence of joint geometrical and mechanical properties on the stability of underground excavations. Shen and Barton (1997) investigated the effect of joint spacing on the size and shape of the disturbed zone around a tunnel; as the joint spacing decreases, blocks could be observed falling into the tunnel while the difference between the two models with small joint spacing was minor; the same phenomena could be obtained for the open zone where joints are open. However, distinct increase of the shear zone (large shear displacement) can be seen with the decrease of joint spacing. Additionally, the varying joint orientation changed the shape of the failure area around the tunnel. Bhasin and Hoeg (1997) performed a sensitivity analysis on joint spacing and joint strength parameters. They found the rotational shear deformation is likely to occur for the large size blocks, while inside highly fractured rock masses the stresses were built up, resulting in smaller displacements instead. With respect to the joint strength parameters, an insignificant difference existed in deformations around the opening by varying joint roughness coefficient and joint residual friction angle, however, a great reduction of displacements was caused by a larger joint wall compressive strength. Hao and Azzam (2005) assessed the stability of an underground cavern by considering the influence of the dip angle, shear strength and location of a fault. According to the results, a critical dip angle existed where the plastic zones and maximum displacement around the opening reached the peak values, and the critical dip changed with the location of the fault with respect to the excavation. The existence of fault resulted in the asymmetric distribution of plastic zones and deformations around the opening; the most unstable case seemed to be the one where the fault intersected with the right wall and caused sliding blocks. Results also showed that a higher fault friction angle effectively reduced the rock mass failure and displacement and their asymmetrical distributions; besides, the critical fault dips were slightly changed. In the study of Wang et al. (2012), the effects of joint stiffnesses and joint friction angle on the tunnel displacements were evaluated; results showed that joint shear stiffness was more sensitive than the joint friction angle on tunnel displacements.

Additional investigations pertinent to the discontinuities can also be found on the selection of appropriate joint constitutive laws (Souley et al., 1997; Cui et al., 2016; Sainsbury and Sainsbury, 2017). So far, efforts have still been making by numerous researchers to seek the realistic representation or modeling of discontinuities for practical problems.

4.3 Equivalent continuum modeling of rock masses

Although the fully explicit representation of rock joints can be achieved using modern numerical techniques for small domain size problems, the equivalent continuum approach still enjoys wide applications in cases where sufficient information of discontinuities is unavailable or that the problem domain is large and rather complex. A discontinuum-equivalent continuum approach has been applied by some researchers (Wu and Kulatilake, 2012a, 2012b; Wang et al., 2012; Kulatilake and Shu, 2015; Sherizadeh and Kulatilake, 2016, 2016; Huang et al., 2017), where the large-scale discontinuities were explicitly modeled while the effect of the small-scale discontinuity properties were implicitly incorporated by combining with the intact rock properties. In equivalent continuum modeling, the mechanical properties of jointed rock masses, which contain the effect of small-scale discontinuities, have to be estimated by using direct or indirect methods.

Direct methods include laboratory and in-situ tests. To obtain the realistic property values for a jointed rock mass, the rock sample should contain enough discontinuity features to represent the overall behavior, which is difficult to be implemented in the laboratory. A variety of in-situ tests, such as compression, shear, plate-bearing, flat jack, borehole jacking and dilatometer tests, have been available to measure the deformability modulus and the strength of rock masses (Bieniawski and Van Heerden, 1975; Bieniawski, 1978). However, the results of in-situ tests are site-dependent and include many uncertainties due to insufficient information on the discontinuity system and their properties of the tested rock mass; the implementation is normally difficult, time-consuming, and expensive.

Under such conditions, indirect methods such as empirical relations, analytical method and numerical method can be used (Kulatilake et al., 1992). Bieniawski (1978) compiled the results of in-situ tests for a number of projects and found a variety of uncertainties in the obtained data; it was suggested that the in-situ tests should be performed with two or more methods for cross-checking; by correlating the in-situ results with the RMR values, an estimate of the rock mass modulus was obtained. Further investigations were made by Serafim and Pereira (1983), Grimstad and Barton (1993), Barton (2002), Hoek and Brown (1997), and Hoek and Diederichs (2006) to relate the rock mass modulus with RMR, Q and GSI systems. Similarly, empirical equations have been also derived to determine the strength of rock masses by scaling down the intact rock strength based on geological conditions (Hoek and Brown, 1980; Kalamaris and Bieniawski, 1995; Barton, 2002; Hoek et al., 2002). For instance, Hoek (1994) and Hoek et al. (1995) introduced the GSI system to account for the reduction of rock mass strength in different geological conditions and was used to estimate the strength parameters. Hoek et al. (2002) proposed the equations to determine the Mohr-Coulomb parameters c and f within certain ranges of the confining stress for tunnels and slopes.

Analytical solutions have been given by some researchers to take into account the simple fracture patterns (Singh, 1973; Amadei and Goodman, 1981; Stephansson, 1981; Gerrard, 1982; Fossum, 1985; Hu and Huang, 1993). The global deformation moduli or compliances used for the analysis were obtained by superimposing the compliance of intact rock and that of the fractures. In these analyses, the geometry of the fracture system was assumed to be regular and persistent; the interaction between the joints was not considered.

The numerical method, which takes advantage of the high efficiency of computer technology, is another indirect way to estimate the rock mass properties. Kulatilake (1985) introduced a procedure to determine the elastic constants and strength of different-sized blocks using the finite element analysis. The representative elementary volume (REV) is a certain minimum volume over which the rock mass properties may not change significantly with respect to the effect of fractures. By generating realistic fracture systems, Wang et al. (2002) and Wu and Kulatilake (2012a) obtained the sizes and property values of the REV for their models. In the subsequent study of Wu and Kulatilake (2012b), the REV properties were used to represent the combined rock mass properties of intact rock and four sets of minor discontinuities; the continuum/discontinuum stress analysis was performed by incorporating the major discontinuities explicitly. In addition, the back-analysis based on the field measurements also provides a way to estimate the in-situ properties of rock masses (Sakurai and Takeuchi, 1983; Cai et al., 2007b; Shreedharan and Kulatilake, 2016).

Another essential aspect of the equivalent continuum modeling is to select the appropriate constitutive relations for the rock masses. According to the practical experience of Hoek and Brown (1997), the post-peak characteristics of rock masses vary with the rock quality. For very good quality hard rock masses, the post-peak failure normally appears as brittle. For such rock masses, the strength of the rock mass abruptly drops to the residual value. The average quality rock masses, however, exhibit a gradual strength loss in the post-failure region, corresponding to the strain softening behavior. For very poor quality rock masses, the post-peak behavior can be represented typically as perfectly plastic which means the rock mass strength stays constant after yielding. The triaxial test data on marbles obtained by Wawersik and Fairhurst (1970) revealed that the post-failure behavior of the rock changes from brittle to ductile with the increasing confining stress. Due to the low confinement around the excavation boundaries, the rock masses are subjected to the unfavorable softening scenario. In the article of Egger (2000), the importance of softening rate on the tunnel stability was stressed; he also discussed the needed support pressure for preserving the rock mass strength at the perfectly-plastic level. The role that post-failure behavior of rock masses plays on the failed or plastic region and the deformations around underground excavations have been investigated by many other researchers (Hoek and Brown, 1997; Cai et al., 2007a; Alejano et al., 2009; Wang et al., 2011; Alejano et al., 2012). The estimation of the post-failure parameters can be achieved by the direct measurements at the site or in the laboratory. The large-scale tests in the field are too expensive to conduct. For most cases when the tests are not available, indirect approaches are generally used, such as the numerical back-analysis conducted by Crowder and Bawden (2006), and the estimation based on the GSI system (Cai et al., 2007a).

4.4   Effects of in-situ stress and construction sequence

The initial stress in rocks is normally described by three components: a vertical stress which is mainly caused by the overburden weight and two horizontal components which are dominated mainly by the tectonic stresses; the horizontal stresses could be larger or smaller than the vertical stress in a wide range (Stille and Palmström, 2008; Stephansson and Zang, 2012). To determine the in-situ stresses, field measurements, such as hydraulic fracturing, sleeve fracturing, borehole breakout, borehole relief methods and over-coring are available. The World Stress Map (WSM), which is a global compilation of the information of the current stress field of the Earth's crust, offers the orientations rather than the magnitudes of tectonic stresses (Zang et al., 2012). Because of the large-scale and limited data points, the WSM may be used as a preliminary check or consultation of the stress state for a region. The magnitude and orientation of the regional stresses are complicated and may vary due to the topographical effect, geological unconformities, stratification and geological structures such as faults, dikes, joints and folds (Stephansson and Zang, 2012; Tan et al., 2014a, 2014b). Stephansson and Zang (2012) illustrated the influence of lithology and structures on the distributions of in-situ stresses and suggested that the numerical stress modeling was a useful aid to predict the overall stress state for an area with great variations and uncertainties. Tan et al. (2014a, 2014b) investigated the influence of inclined strata and faults on the in-situ stress state for an open-pit mine; they concluded that to estimate the in-situ stresses for a numerical model having a complex geological system, the appropriate way is to perform stress analysis on this system with proper boundary stresses.

Underground excavation disturbs the initial stress state in rock masses and causes stress redistributions. At low in-situ stress levels, the rock mass tends to be unstable due to the low normal stress acting on the joints (Stille and Palmström, 2008). As in-situ stress increases with depth, the tangential stresses around the opening periphery would set up, of which the magnitudes depend upon the initial stress state and upon the excavation shape. Instabilities may arise once the stress is greater than the strength of rock masses. For competent rocks, failures such as buckling, rupturing, slabbing and rock burst are likely to occur, while the plastic behavior and squeezing are normally observed in the incompetent or weak rock masses (Bhasin and Grimstad, 1996; Palmström and Stille, 2007). As an important factor affecting the stability of underground excavations, the impact of in-situ stress on the failure mechanism and displacements have been studied extensively.

Hoek and Brown (1980) provided the maximum stress values around the excavations with various excavation shapes and horizontal to vertical in-situ stress ratios (lateral stress ratios) for isotropic rock masses under elastic condition; results showed that the maximum stress in the roof increased with the increasing lateral stress ratio, while the maximum stress in the sidewall decreased with increasing lateral stress ratio. A 90-degree rotation in the principal stress state around a square tunnel was observed as the lateral stress ratio changed from 0.5 to 2.0 while the magnitude of the major principal stress was doubled. Martin et al. (1999) suggested that the brittle failure in massive rocks around the underground opening was initialized when the ratio of the maximum tangential boundary stress to the laboratory unconfined compressive strength exceeds 0.4. They also proposed an equation expressed with Hoek-Brown parameters to estimate the depth of the stress-induced failure; as the ratio of the maximum principal stress to the minimum principal stress increased, the shape of the damage region enlarged from the localized point to a notched area. Jia and Tang (2008) studied the influence of lateral stress ratio on the failure modes and deformations around a tunnel. For low lateral stress ratio cases, cracks were observed in the sidewalls of the tunnel; flexure of the rock mass and sliding along the joints were seen. For high lateral in-situ stresses, the failures, however, concentrated on the roofs and floors. As the lateral stress ratio increases, deformations on the sidewalls were increased but decreased on the roof and floor. Using numerical results for an underground powerhouse under various rock mass and geological conditions, Zhu et al. (2004) derived a few empirical equations to predict the displacement at a key point on the sidewall, which increased with the rock mass modulus, overburden thickness and excavation height as well as the lateral stress ratio. Eberhardt (2001) investigated the stress changes around a tunnel with the advancing excavation. The changes of stress magnitudes and directions at the fixed roof and wall locations were addressed in detail; the variation patterns changed with the different initial stress field. The author also pointed out that the brittle fracture failure which begins with the initiation and propagation of microfractures is closely related to the magnitudes and orientations of the excavation-induced principal stresses. By performing the equivalent continuum/discontinuum modeling, Wu and Kulatilake (2012b) found that with increasing lateral stress ratio, the maximum horizontal displacements on the ribs of tunnels increased while the maximum vertical displacements on the roof and floor decreased. In the study of Sherizadeh and Kulatilake (2016), the impacts of the orientation and magnitude of horizontal in-situ stress on the roof stability of underground excavations were evaluated; conclusions were made that when the maximum lateral stress ratio increased from 1 to 2, the increases of vertical displacements and shear failure on the roof were observed. As the maximum horizontal stress orientation varied, the location and severity of roof failure as well as that of roof displacements were changed.

Stress redistribution around the excavations, on the other hand, depends on the construction methods (i.e. excavation method and supporting sequence) and hence influence the response of rock masses. In practice, the implemented excavation and supporting procedures are diverse, depending on the factors such as excavation geometry, excavation shape, rock mass condition and so forth (Kim et al., 1999; Barla, 2001). For example, the conventional construction methods for tunneling in squeezing rock condition include the side drift method, the top heading and benching down excavation method, as well as the full-face excavation method; each excavation method may be associated with a suitable supporting strategy (Barla, 2001). Cai (2008) pointed out the difference between the two computer codes, FLAC and Phase^2, in simulating the tunnel excavation process, and stressed the influence of the stress paths on the plastic yielding zone distributions. Simulation of the realistic excavation and supporting sequences that implemented in the field is essential for accurate prediction of the rock mass behavior (Cai, 2008; Cantieni and Anagnostou, 2009).

Since the stress path of underground excavation or tunneling is three-dimensional (Eberhardt, 2001; Kaiser et al., 2001; Cai, 2008; Cantieni and Anagnostou, 2009), it should be included in the analysis. For 2-D modeling, the necessary 3-D effects of excavation include the pre-face response, the displacement and plasticity at the face, and the subsequent development of deformation and yielding. Based on the assumption that tunnel boundary deforms progressively as the tunnel face passes the model section, these effects can be incorporated by using the methods such as field stress vector/average pressure reduction, excavation of concentric rings, and face replacement or destressing (Vlachopoulos and Diederichs, 2014). In the first approach, an internal pressure/traction equal to in-situ stress is applied to excavation boundary, and then the monotonic reduction of the pressure/traction is assumed to represent the excavating process (Vlachopoulos and Diederichs, 2014). This excavation relaxation method was applied by Zhao and Cai (2010) to simulate the full-face excavation in a 2-D analysis. However, these methods are applicable only for the simplified ideal situations. For real-world cases, the behavior of tunnel boundary is more complicated than gradually deformed. For instance, if multiple excavations were made, or a large excavation was driven in stages, the influence of close-by excavations or of staged excavations should be considered. Under such conditions, the above methods are difficult to apply, and 3-D analysis is required.

To precisely simulate the delayed installation of rock supports, the deformation occurred prior to applying support needs to be considered. It can be estimated using the analytical methods for simplified cases, that is by constructing the longitudinal displacement profile in the convergence-confinement method or using the numerical back-analysis for complicated conditions (Barla, 2001). For example, based on the results of a 3-D numerical simulation, Janin et al. (2015) back-calculated the stress release coefficients, λ_1 and λ_2, respectively used for the installation of tunnel lining and tunnel invert in the 2-D analysis. Vardakos et al. (2007) and Shreedharan and Kulatilake (2016) adopted the stress relaxation method by applying the interior stresses on the excavation boundaries with certain stress loss levels, which correspond to the timing of support installation. In the former study (Vardakos et al., 2007), the stress loss factor was assumed as 0.5 for the rock bolts and as 0.65 for the shotcrete; the latter authors (Shreedharan and Kulatilake, 2016) tried multiple stress loss levels at every 10% gap. It should be mentioned that the stress relaxation method used here and that mentioned in the above paragraph have different physical meanings. Previously, those were applied to simplify the 3-D effects of excavation into 2-D cases. In the current paragraph, they are utilized to take into account the deformation prior to or the time of support installation.

4.5 Summary

The rock mass contains pre-existing defects, which could significantly weaken the rock mass strength, increase the rock mass deformability, and make the rock mass behavior complicated. Therefore, the incorporation of discontinuities or of their effects is necessary for the numerical modeling of rock mass stability of underground excavations. The joint geometry patterns and joint material properties are two crucial issues. Although several means have been developed for the generation of fracture systems from mapping data, most of the developed models are probably restricted to simple stress analysis or the small-scale problems.

Estimation of the joint material properties is a difficult task because of the existence of scale effect, the uncertainties in the joint condition, the joint filling material and so forth. Parametric studies hence have been performed extensively with respect to the size, density, spatial distribution and so forth of the discontinuities.

Because the complex fracture system is mostly difficult to simulate completely and also too large to represent in a numerical model, the equivalent continuum modeling is preferred, where the rock mass property and the constitutive behavior are of great importance. The available approaches for the estimation of rock mass properties include the direct (i.e. in-situ and laboratory tests) and the indirect methods (i.e. the analytical, empirical and numerical methods). Generally, in-situ tests are time-consuming, expensive, site dependent and lack data on existing fracture system in the tested block; scale effect is one of the major limitations of laboratory tests. The analytical method yet seems to be applicable only for the simplified situation. Empirical relations have been developed to correlate the rock mass properties with the rock mass classification indices such as RMR, Q and GSI values. It is an easy and efficient method and has gained wide applications. However, many shortcomings exist. Numerical modeling is another competent method, where the REV properties can be estimated by incorporating the actual fracture system. Because the softening post-failure behavior of rock masses greatly impacts the stability of underground excavations, it should be taken into account in the analysis.

Underground excavation breaks the balance of initial stress state in rocks, causing stress redistributions around the openings. The deformations and failures of surrounding rock masses highly depend on the direction and magnitude of in-situ stresses. Moreover, mining-induced stress change is a three-dimensional problem, which could be affected by the construction method and the rock support sequences. Although a couple of methods were proposed to incorporate the 3-D effects in the 2-D analysis, they were based on the assumptions for idealized situations. The delayed installation of supports has been considered by using the stress relaxation method.

The purpose of rock engineering modeling is to develop an adequate model to capture the most critical factors so that the rock mass behavior can be correctly predicted, providing guidelines for support design or excavation rearrangement. The fundamental and essential aspects for the modeling of rock mass stability of underground structures are presented in previous sections and summarized as above. Generally, each of these factors has been investigated independently and to some extent in the literature. A three-dimensional stress analysis which can incorporate multiple factors is, however, rare to find.

Chapter 5

Theory and background of three-dimensional distinct element code

5.1 Introduction

For the present case study, the 3-D Distinct Element Code (3DEC) is selected as the numerical analysis method. Some aspects of the theory and background of 3DEC will be introduced in this chapter. As mentioned in Chapter 3, the distinct element method is an explicit DEM to model the discontinuous systems. The discontinuities in rock masses are explicitly represented and assigned with constitutive laws. Cundall (1988) and Hart et al. (1988) proposed the numerical formulations of the three-dimensional code (3DEC), where the contacts were treated as deformable, and the blocks can be modeled as both rigid and deformable. In Section 5.2, the scheme of contact detection is first introduced. Subsequently, the mechanical calculation cycle is briefly described. Section 5.3 discusses the discretization techniques available in 3DEC. Following that are the formulations of the block and joint constitutive models and the rock support model relevant to this study.

5.2 Scheme of contact detection

As an essential characteristic of the discrete element code, the automatic recognition of new contacts needs to be realized. Unlike the two-dimensional situation, the identification and detection of contacts in a 3-D system composed of polyhedral blocks are more difficult due to the various contact types (i.e. vertex to vertex, vertex to edge, edge to face and face to face). A robust and rapid scheme was developed by Cundall (1988) to simplify the detection calculations. This scheme involves the identification of whether the blocks are touching, the determination of the maximum gap between the blocks if not touching, the contact type and the direction of potential sliding (Itasca, 2007).

In the scheme, a common plane (c-p) has been introduced, located in the middle of the two blocks (Figure 5.1) (Itasca, 2007). The utilization of c-p largely simplifies and speeds up the contact testing procedures. Only the vertex-to-plane tests are needed; the number of tests is significantly cut down. The plane is kept at the maximum distance from each block if the blocks are not touching; the gap between the blocks can be determined by adding the block-to-plane distances. If the blocks are overlapping (penetrating), the position of the plane is determined by minimizing the distance between the c-p and the vertex of blocks.

Figure 5.2 shows some examples of the common plane between two blocks in two dimensions; the c-p could be touched by different components of the blocks (vertex, edge or face). The unit normal vector of the contact, which defines the plane along which sliding can occur, agrees with the unit normal of the c-p. Once the gap between two blocks is equal to or less than a preset tolerance, a contact is created. On the contrary,

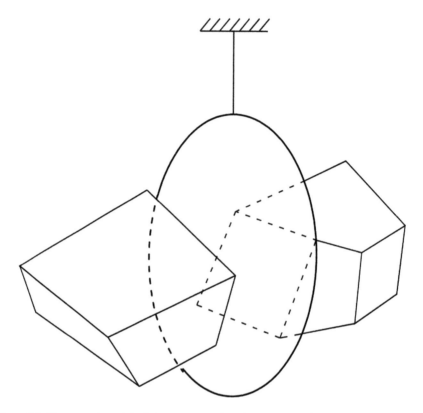

Figure 5.1 Conceptual model of the common plane between two blocks
Source: From Itasca (2007).

if an existing contact is separated more than the tolerance, the contact is deleted. If the two blocks touch the c-p, they are touching each other; otherwise they are not. The contact type can be simply determined by counting the number of vertices of each block touching the c-p, as listed in Table 5.1 (Itasca, 2007).

In 3DEC, the face-to-face contacts are treated as "joints," and the stress-displacement law is assigned to describe their mechanical behavior. If these contacts are detected, they are automatically discretized into sub-contacts. For rigid blocks, the sub-contacts are generally created at the vertices of the block surface. For deformable blocks, the sub-contacts are the nodes of the triangular faces of tetrahedral zones. Assigned with an area and following the joint constitutive relations, the sub-contacts are normally used to transfer the interaction forces between blocks and to deal with sliding and separation (Itasca, 2007).

5.3   Mechanical calculation cycle

In 3DEC, the block system is controlled by the law of motion. Based on the dynamic algorithm, the equations are solved using the explicit finite difference method. For rigid blocks, the translational and rotational velocities (or displacement increments) of block centroid are

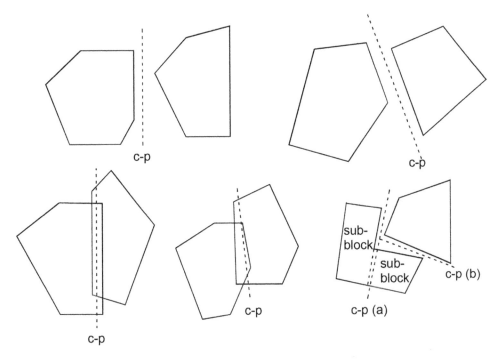

c-p

c-p

c-p

sub-block

sub-block

c-p

c-p (a)

c-p (b)

c-p

Figure 5.2 2-D examples of the common plane between two blocks
Source: From Itasca (2007).

Table 5.1 Determination of the contact type[a]

Number of vertices touching the c-p		Contact type
Block A	Block B	
0	0	null
1	1	vertex-vertex
1	2	vertex-edge
1	>2	vertex-face
2	1	edge-vertex
2	2	edge-edge
2	>2	edge-face
>2	1	face-vertex
>2	2	face-edge
>2	>2	face-face

[a] From Itasca (2007).

calculated at each time step. The deformable blocks, on the other hand, are discretized into finite-difference tetrahedral elements. The velocities (or displacement increments) of grid points (vertices of the element) are calculated, and the constitutive relations are used to update the strains and stresses of the element. After obtaining the contact velocities (or displacement increments) from block motion, the sub-contact forces are determined using the

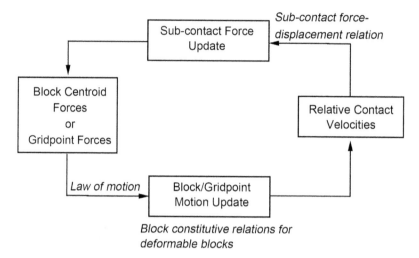

Figure 5.3 Mechanical calculation cycle in 3DEC
Source: Slightly modified from Itasca (2007).

sub-contact force-deformation relation and hence applied to the next calculation step. Figure 5.3 presents the described calculation procedures (Itasca, 2007).

5.4 Discretization

Deformable blocks in 3DEC are internally discretized into tetrahedral (constant-strain) elements (Itasca, 2007). The use of tetrahedral elements can eliminate the occurrence of hourglass modes of deformation (combinations of nodal displacements that produce no strain in polyhedral with more than four nodes) and has the ability of meshing irregular block geometries (Itasca, 2007). However, the typical problem is encountered of accurately modeling of plastic behavior, called "mesh-locking" or "excessively stiff," which is caused by the incompressibility condition of plastic flow (Nagtegaal et al., 1974). This problem can be resolved by the employment of the technique of "mixed discretization" (Marti and Cundall, 1982). In 3DEC, it involves an assemblage of several tetrahedra into quadrilateral/hexahedra (Itasca, 2007). The individual constant-strain tetrahedral elements still experience the incompressible plastic flow, while the volumetric flexibility is increased by evaluating the volumetric behavior over the zones (quadrilateral or hexahedra). The mixed discretization in 3DEC (the GEN quad command) is yet restricted to the six-sided polyhedra.

Nodal mixed discretization (NMD) is a variation of the mixed discretization scheme, where the averaging of the volumetric behavior is performed on nodes instead of zones (Itasca, 2007). It is suitable for the plastic analysis with the capability of meshing complex block geometries. Hence, the NMD is applied in this study.

In the NMD technique, the strain rate ($\dot{\varepsilon}_{ij}$) is divided into deviatoric ($\dot{e}_{ij}$) and volumetric ($\dot{e}$) components as

$$\dot{\varepsilon}_{ij} = \dot{e}_{ij} + \dot{e}\delta_{ij} \tag{5.1}$$

where δ_{ij} is the Kronecker delta function.

The average volumetric strain rate of a tetrahedral element is calculated by the following equation (Eq. 5.2). Then the total strain rate in Eq. (5.1) is updated by substituting the $\dot{e}$ by $\bar{\dot{e}}$ (Itasca, 2007)

$$\bar{\dot{e}} = \frac{1}{4} \sum_{n=1}^{4} \left(\frac{\sum_{k=1}^{m_n} \dot{e}_k V_k}{\sum_{k=1}^{m_n} V_k} \right)_n \tag{5.2}$$

where "4" represents the four nodes of the tetrahedral element, m_n is the number of elements surrounding the element node n, and V_k is the volume of element k.

Higher-order tetrahedral elements (i.e. the 10-node tetrahedra) are also available for the block plasticity problems (the GEN hotetra command) in 3DEC. However, they are suitable to be applied for the situation where the block stress distribution or the block yielding behavior is important. For the problems where the behavior of discontinuities is critical, the higher-order approach seems less appropriate due to the increased complexities involved in contact calculations (Itasca, 2007).

5.5   Block and joint constitutive models

5.5.1   *Mohr-Coulomb plasticity model*

The Mohr-Coulomb (M-C) plasticity model in 3DEC is a conventional constitutive model to represent the shear failure of soils and rocks. In this model, the mechanical behavior of deformable blocks is prescribed by a linear-elastic, perfectly plastic relation (Figure 5.4) with the Mohr-Coulomb failure criterion ($f_s = 0$), including a tension cutoff ($f_t = 0$)

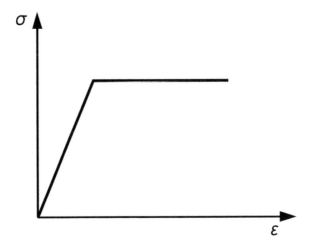

Figure 5.4 Typical stress-strain curve of the Mohr-Coulomb model

Source: Reprinted by permission from Springer Nature Customer Service Centre GmbH: Springer, *Geotechnical and Geological Engineering*, "Rock mass stability investigation around tunnels in an underground mine in USA," Y. Xing, P.H.S.W. Kulatilake & L. A. Sandbak, 2017.

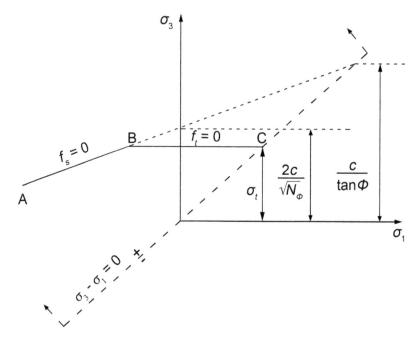

Figure 5.5 Mohr-Coulomb failure criterion in 3DEC

Source: From Itasca (2007).

(Itasca, 2007). The expressions of f_s and f_t are given by

$$f_s = \sigma_1 - \sigma_3 N_\phi + 2c\sqrt{N_\phi} \qquad (5.3)$$

$$f_t = \sigma_3 - \sigma_t \qquad (5.4)$$

where σ_1 is the major principal stress, σ_3 is the minor principal stress, ϕ is the friction angle, c is the cohesion, σ_t is the tensile strength and

$$N_\phi = (1 + \sin \phi)/(1 - \sin \phi) \qquad (5.5)$$

The shear failure is detected if $f_s < 0$ and the tensile failure occurs when $f_t > 0$ (compressive stresses are negative). Figure 5.5 shows the failure envelope. The maximum tensile strength is expressed as

$$\sigma_{tmax} = \frac{c}{\tan \phi} \qquad (5.6)$$

5.5.2 *Strain-hardening/softening model*

The strain-hardening/softening model in 3DEC is an extended block constitutive model based on the M-C failure criterion. Being different from M-C model, the strain-hardening/

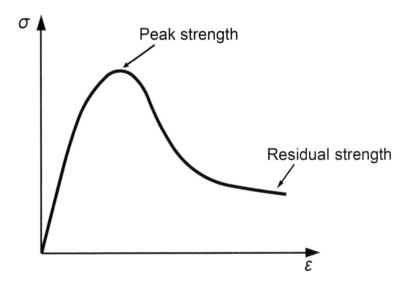

Figure 5.6 Typical stress-strain curve for strain-softening behavior

Source: Reprinted by permission from Springer Nature Customer Service Centre GmbH: Springer, *Geotechnical and Geological Engineering*, "Rock mass stability investigation around tunnels in an underground mine in USA," Y. Xing, P.H.S.W. Kulatilake & L. A. Sandbak, 2017.

softening model is able to describe the non-linear behaviors (i.e. softening and hardening) after peak failure. Figure 5.6 shows the typical stress-strain curve of the strain-softening material. In this model, the softening or hardening behaviors, in terms of the variations of M-C parameters (cohesion, friction angle, dilation, tensile strength), can be expressed as functions of the deviatoric plastic strain (Itasca, 2007). Figure 5.7 presents the conceptual models of the softening variations of the strength parameters, cohesion, friction angle, and tensile strength. They are approximated as linear segments in 3DEC. The e^{ps} and e^{pt} are the deviatoric plastic shear and tensile strains, respectively, of which the increments are defined by Eqs. (5.7) and (5.8) (Itasca, 2007).

$$\Delta e^{ps} = \frac{1}{\sqrt{2}}\sqrt{(\Delta\varepsilon_1^{ps} - \Delta\varepsilon_m^{ps})^2 + (\Delta\varepsilon_m^{ps})^2 + (\Delta\varepsilon_3^{ps} - \Delta\varepsilon_m^{ps})^2} \qquad (5.7)$$

$$\Delta e^{pt} = |\Delta\varepsilon_3^{pt}| \qquad (5.8)$$

where $\Delta\varepsilon_1^{ps}$, $\Delta\varepsilon_3^{ps}$, and $\Delta\varepsilon_3^{pt}$ are respectively the plastic-strain increments and can be obtained based on the flow rule; $\Delta\varepsilon_m^{ps}$ is the volumetric plastic shear strain increment, given by

$$\Delta\varepsilon_m^{ps} = \frac{1}{3}(\Delta\varepsilon_1^{ps} + \Delta\varepsilon_3^{ps}) \qquad (5.9)$$

5.5.3 Coulomb-slip joint model

The Coulomb-slip joint model is a basic constitutive model in 3DEC used for joints (Itasca, 2007). Constant joint stiffnesses, K_n and K_s, are specified to describe the linear behavior of

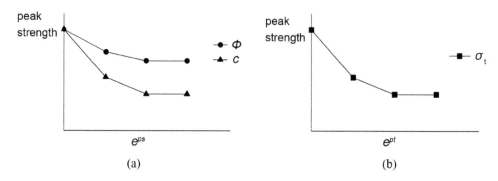

Figure 5.7 Softening behavior of strength parameters with deviatoric plastic strain: (a) cohesion (c) and friction angle (ϕ); (b) tensile strength (σ_t)

Source: Modified from Itasca (2007). Reprinted by permission from Springer Nature Customer Service Centre GmbH: Springer, *Geotechnical and Geological Engineering*, "Rock mass stability investigation around tunnels in an underground mine in USA," Y. Xing, P.H.S.W. Kulatilake & L. A. Sandbak, 2017.

joints prior to failure. The maximum tensile and shear forces, T_{max} and F^s_{max}, of joints are respectively given by (Itasca, 2007)

$$T_{max} = -TA_c \qquad (5.10)$$

$$F^s_{max} = c_j \cdot A_c + F^n \tan \phi_j \qquad (5.11)$$

where T is the joint tensile strength, A_c is the area of the contact (sub-contact), c_j is the joint cohesion, F^n is the joint normal force, and ϕ_j is the joint friction angle.

Once failure occurs (the maximum forces are reached) in either tension or shear, the T_{max} and c_j are set to zero. For shear failure, the joint shear force decreases and stays constant at $F^n \tan \phi_j$. For tensile failure, both the joint normal and shear forces reduce to zero.

5.5.4 Continuously yielding joint model

Continuously yielding joint model in 3DEC can simulate more realistic behaviors of joints, such as joint shearing damage, normal stiffness dependence on normal stress and decrease in the dilation angle with plastic shear displacement (Itasca, 2007).

In this model, the joint normal and shear stress increments are obtained by (Itasca, 2007)

$$\Delta\sigma_n = K_n \Delta u_n \qquad (5.12)$$

$$\Delta\tau = FK_s \Delta u_s \qquad (5.13)$$

where σ_n is the joint normal stress; τ is the joint shear stress; u_n and u_s are the joint normal and shear displacements; F is a factor related to the stress history, current stress state, joint peak and basic friction angle, and the joint roughness parameter; and K_n and K_s are joint normal and shear stiffnesses that can be expressed as functions of the normal stress according to Eqs. (5.14) and (5.15):

$$K_n = a_n \sigma_n^{e_n} \qquad (5.14)$$

$$K_s = a_s \sigma_n^{e_s} \tag{5.15}$$

where a_n and a_s are the coefficients of the equations; and e_n and e_s are the joint stiffness exponents.

The strength equation for this joint model is given by

$$\tau = \sigma_n \tan(\phi_j + i) \tag{5.16}$$

where ϕ_j is the joint basic friction angle and i is dilation angle.

5.6   Cable structural model

The cable structural model is one of the available models for simulating rock bolts in 3DEC. It provides the restraint along the length for the surrounding rock mass to support itself. The support structure is assumed as a one-dimensional element and divided into several segments with nodal points at each segment end. The mass of each segment is lumped at the nodal point. Figure 5.8 shows the conceptual mechanical model of the structure (Itasca, 2007). The axial stiffness of the steel is represented by the spring between the two adjacent nodes; the bond resistance, offered between the grout and steel and between the grout and

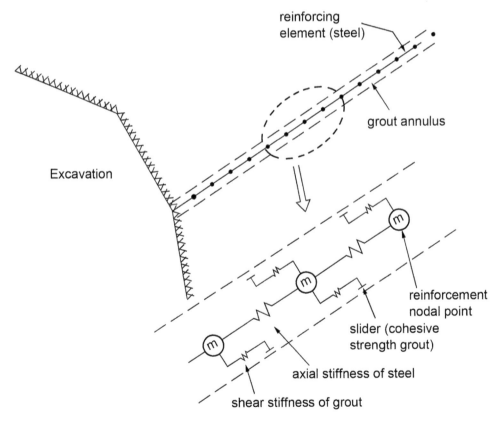

Figure 5.8 Conceptual mechanical representation of the cable element reinforcement

Source: From Itasca (2007).

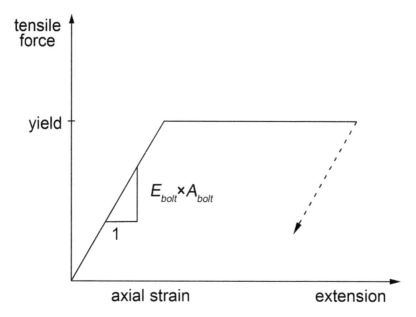

Figure 5.9 Reinforcement axial behavior of the cable element in 3DEC
Source: From Itasca (2007).

rock mass, is represented by the springs and sliders between the structural nodes and the block nodes. The formulations of this support element are illustrated as follows.

For each segment, the axial force increment of the reinforcement, ΔF_c^t, can be determined by the axial displacement increment

$$\Delta F_c^t = -\frac{E_{bolt}A_{bolt}}{L}\Delta u^t \tag{5.17}$$

where E_{bolt} is Young's modulus of the reinforcement, A_{bolt} is the reinforcement cross-sectional area, u^t is the relative axial displacement of the two nodes of the segment and L is the length of the segment.

Tensile failure occurs if the tensile force, F_c^t, reaches the yield limit (tensile yield force) and the maximum tensile force has been set, as shown in Figure 5.9.

The shear behavior of the grout annulus (spring-slider system in Figure 5.8) is described by

$$\frac{F_c^s}{L} = K_{bond}(u_c - u_m) \tag{5.18}$$

where F_c^s is the shear force in the grout, K_{bond} is the grout shear stiffness, u_c is the axial displacement of the reinforcement and u_m is the axial displacement of the rock mass.

The maximum shear force per length, S_{bond}, has been assigned to prescribe the grout (bond) failure, as shown in Figure 5.10.

In general, the material properties of the grout (bond) (i.e. K_{bond} and S_{bond}) can be estimated from the pull-out tests. Under the circumstance where the pull-out test results are

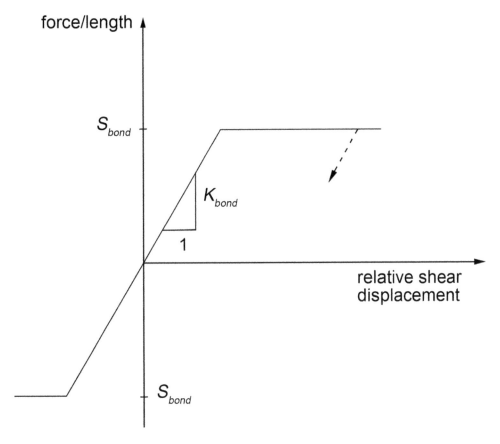

Figure 5.10 : Bond shear behavior of the cable element in 3DEC
Source: From Itasca (2007).

unavailable, Eqs. (5.19) and (5.20) can be used as suggested by the 3DEC manual (Itasca, 2007) to estimate the K_{bond} and S_{bond}.

$$K_{bond} = \frac{2\pi G_g}{10\ln(1 + 2t/D)} \tag{5.19}$$

$$S_{bond} = \pi(D_c + 2t_c)\tau_{peak} \tag{5.20}$$

where G_g is the grout shear modulus, D_c is the reinforcement diameter and t_c is the annulus thickness.

The parameter τ_{peak} in Eq. (5.20) is the peak shear strength of the grout, which is related to the uniaxial compressive strength of the rock and grout and the quality of the bond between the grout and the rock. In the cases where failure takes place at the reinforcement/grout interface, the $(D_c + 2t_c)$ in Eq. (5.20) should be substituted by (D_c).

5.7   Summary

Because of the capabilities of allowing large displacements and rotations of discrete blocks (including the complete detachment), as well as modeling complex constitutive behaviors for both rock blocks and the discontinuities, the 3DEC was chosen to investigate the rock mass stabilities for the case study. This chapter provides some underlying concepts and their corresponding numerical formulations, which include the contact detection, the mechanical calculation cycle, the discretization, the block and joint constitutive models and the rock reinforcement element. They form the basis of understanding the numerical code and the theories of numerical simulations in Chapter 7.

Chapter 6

Conducted laboratory tests and results

6.1 Introduction

Rock samples were collected by the mining company and sent to the Geomechanics Laboratory at the University of Arizona for testing. Figure 6.1 shows the boreholes (short and straight lines) where the rock cores were taken with respect to the tunnel system. Diverse locations and orientations of the boreholes can be observed. Detailed information about the locations and geometry of these boreholes is given in Table 6.1; the location ranges of the selected study area are provided after the table. The involved rock types include dacite, mudstone and limestone. Geomechanical laboratory tests were carried out, including Brazilian, uniaxial compression, ultrasonic, triaxial, uniaxial joint compression and small-scale direct shear tests. The tests were performed as per the American Society for Testing and

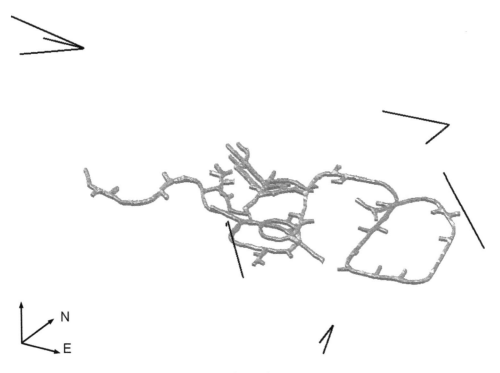

Figure 6.1 Locations of the boreholes and tunnel system

Table 6.1 Locations and orientations of the boreholes (from the mining company)

Borehole No.	Collar location			Trend (°)	Plunge (°)	Length of the hole (m)
	East (ft)	North (ft)	Elevation (ft)			
TUO 2464A	68,121	59,778	3426	162.1	85.5	4.6
TUO 2466	68,123	59,789	3428	96.5	74.7	89.9
TUO 2504	68,391	61,492	2751	291.0	23.8	141.7
TUO 2509	68,397	61,496	2749	309.5	50.5	120.1
TUO 2866	68,710	59,635	3169	258.0	75.0	53.3
TUO 2867	68,709	59,635	3169	258.5	61.0	38.1
TUO 2596	66,810	61,050	2954	283.0	16.0	73.2
TUO 2599B	66,809	61,047	2958	263.5	−14.0	106.7
TUO 2882	68,514	61,192	2753	132.0	7.0	143.3
TUO 2901	66,807	61,046	2953	261.0	19.0	94.5

Note: The ranges of the selected study area in the three perpendicular directions: E67800–E68200 (122 m); N59900–N60300 (122 m); Elevation: 2975–3375 ft (122 m).

Material (ASTM) standards. A detailed description of the sample preparation and the results and analyses of the tests is presented in this chapter.

6.2 Tested rock materials

As mentioned in Chapter 2, the interest area is located in the stratigraphic unit OC5, which mainly consists of the limestone with thinly laminated to thinly bedded mudstones. In addition, the intrusive Main Dacite Dike exists, of which the major rock type is dacite. Therefore, the three types of rocks were prepared for testing, which are respectively LM (intercalated limestone and mudstone with limestone > mudstone), ML (intercalated mudstone and limestone with mudstone > limestone) and dacite. Figure 6.2 presents the rock cores before testing as well as some prepared samples.

6.3 Experimental procedures

The samples from the underground mine were tested with the aim of estimating the tensile strength, uniaxial compressive strength (UCS), Young's modulus, Poisson's ratio, Mohr-Coulomb strength parameters (i.e. friction angle and cohesion) of the intact rock and joint normal and shear stiffnesses and joint friction angle of the smooth joints. The experimental procedures of each test are described as follows.

6.3.1 Brazilian test

The Brazilian test is a simple indirect testing method to obtain the tensile strength of brittle material such as concrete, rock and rock-like materials. In this test, a thin circular disc is diametrically compressed to failure. The indirect tensile strength is typically calculated based on the assumption that failure occurs at the point of maximum tensile stress (i.e. at the center of the disc). The samples were prepared as per ASTM standard D3967 (2008). Using the Allied Powercut 10 saw, the samples were cut with an average value of 0.42 for length to diameter ratio, which lies between 0.2 and 0.75. The diameter and length of the sample were determined by averaging three measurements for each sample.

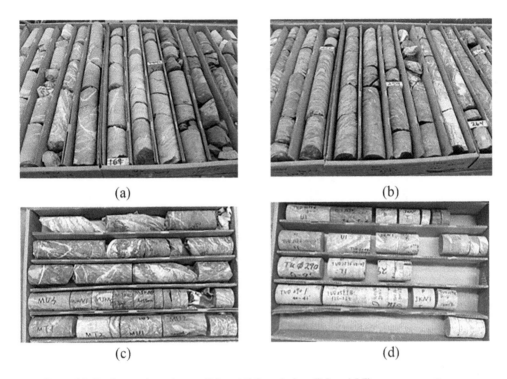

(a)

(b)

(c)

(d)

Figure 6.2 Rock samples prior to ([a] and [b]) and after ([c] and [d]) test preparation

To minimize the influence of the pre-existing fractures, the fractures seen on the sample surface were highlighted and the loading directions were marked to ensure proper orientation in both sides of each sample. A layer of masking tape was wrapped around the edge of the sample to prevent the sample from separating after breakage. Two curved bearing blocks were used to load and reduce contact stresses. The sample was placed into the center of the curved bearing blocks and this whole setup was placed into the Brazilian test device shown in Figure 6.3. The loading rate calculated according to the sample dimensions was entered into the computer. A continuous increasing load was applied until the specimen failed. Finally, the failure load was obtained from the computer data; photos of failed samples were taken to demonstrate the planes of failure (Figure 6.4).

6.3.2 Uniaxial compression test

For the uniaxial compression test, the ASTM standard D7012 (2014) was followed. According to the ASTM standard, the length to width ratio of the sample should not be more than 2.5 or less than 2. The ratio of 2 was used for testing. Chipping on the end surfaces of samples and deviation from parallelism of two end surfaces in each sample are two typical errors that can occur during the cutting, and they lead to errors in the value of the strength obtained in this test. To overcome the potential for these errors, the two end faces of each sample were smoothed by a grinding machine (Figure 6.5), as suggested in ASTM D4543 (2008). For each sample, the diameter and height were measured three times and

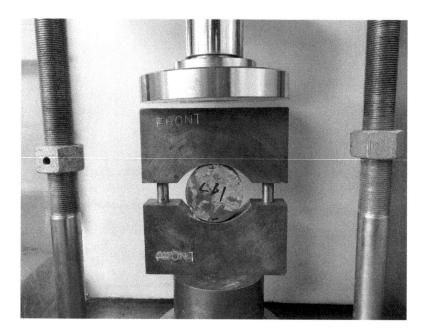

Figure 6.3 Compression testing machine used for Brazilian test

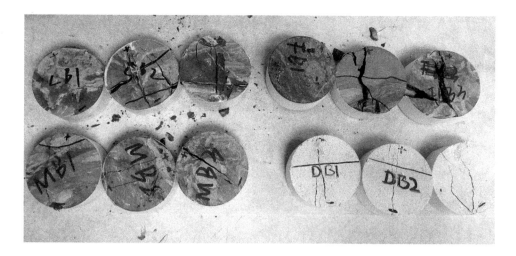

Figure 6.4 Failed samples of Brazilian tests

average values were determined. The weights of the samples were measured and their densities were calculated. Electrician's tape was wrapped around the samples to hold them together and prevent their explosive failure during testing. After placing in the testing apparatus (Figure 6.6), a displacement rate of 0.0076 mm/s was applied for the test.

For measuring Young's modulus (E) and Poisson's ratio (μ), two methods were followed. The first method is to use strain gauges to calculate the axial and lateral strains,

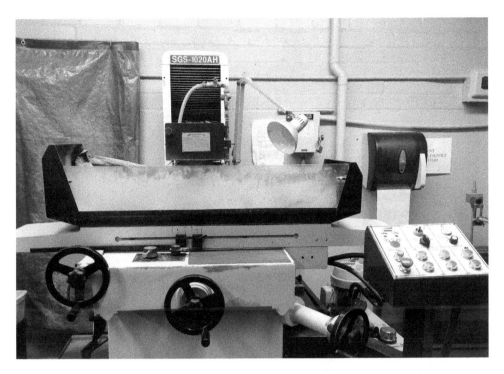

Figure 6.5 Grinding machine used to smoothen the sample faces

and hence to estimate the Young's modulus and Poisson's ratio. Figure 6.7 provides the photos of the samples with attached strain gauges—one vertical and one horizontal for each sample. After installation, the connections were checked to ensure that they formed a complete Wheatstone's bridge circuit. Then the sample was tested in a regular way with a displacement rate of 0.0076 mm/s. After failure, the slope of the linear region of the stress-strain curve was used to estimate the Young's modulus and the ratio of the lateral strain to axial strain was used to obtain the Poisson's ratio. After failure, the photos were taken to identify if the samples failed along a pre-existing discontinuity or in an intact zone or both (Figure 6.8).

6.3.3 *Ultrasonic test*

The second method to determine E and μ is known as the ultrasonic or seismic test. In this method, the P (compression) and S (shear) waves are used to measure the time taken for the wave to cover the length of the specimen to determine the wave velocity and subsequently, the dynamic Young's modulus and Poisson's ratio as described in ASTM standard D2845 (2008). The specimen was positioned horizontally and its cylindrical ends were cleaned and greased to provide cohesion between the rock and transducers. The time taken for the wave pulses to move from one end of the sample to another was recorded on an oscilloscope and the velocity was determined using Newton's laws of motion. After obtaining the wave

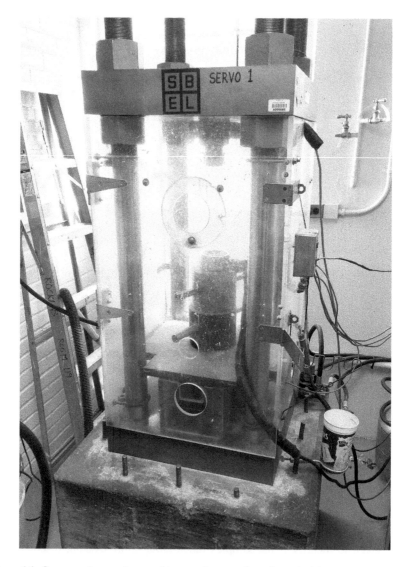

Figure 6.6 Compression testing machine used to conduct the uniaxial compression test

velocities and density of the specimen, Young's modulus (E) and Poisson's ratio (μ) were calculated by Eqs. (6.1) and (6.2).

$$E = [\rho V_s^2 (3V_p^2 - 4V_s^2)]/(V_p^2 - V_s^2) \tag{6.1}$$

$$\mu = (V_p^2 - 2V_s^2)/[2(V_p^2 - V_s^2)] \tag{6.2}$$

where V_p is the propagation velocity of the compression wave, V_s is the propagation velocity of the shear wave and ρ is the density of the rock sample.

Figure 6.7 Uniaxial compression test samples with strain gauges

Figure 6.8 Failed samples of uniaxial compression test

6.3.4 *Triaxial test*

For triaxial tests, the ASTM standard ASTM D7012 (2014) was followed. For some samples, the used length to width ratio was less than 2; for those samples, a correction was used. For each sample, the diameter and height were measured three times and an averaged value for each sample was determined. The samples were first prepared by cutting and grinding (see Figure 6.9) and then were placed in the triaxial cell shown in Figure 6.10 for testing. A manually operated compression pump (see Figure 6.11) was connected to supply compression to the oil inside the cell, and subsequently, the confining stresses were applied by the lateral fluid pressure. The servo tester shown in Figure 6.11 was used to provide the axial load. While keeping the confining stress constant, the axial compressive stress was increased until the rock sample failed. Each triaxial test would get a failure stress and a

Figure 6.9 Prepared triaxial test samples

Figure 6.10 Triaxial cell to place specimen for the test

Figure 6.11 Compression testing machine and compression pump used for triaxial test

confining stress. The Mohr-Coulomb strength parameters (i.e. friction angle and cohesion) of the intact rock were calculated through the obtained relation between the principal stresses σ_1 and σ_3 or Mohr-Coulomb circles. Figure 6.12 shows the failed samples of the triaxial tests.

Figure 6.12 Failed samples of triaxial tests

6.3.5　Uniaxial joint compression test

According to Goodman et al. (1968), joint stiffnesses can be defined by

$$d\sigma_n = K_n du_n \text{ and } d\tau = K_s du_s \qquad (6.3)$$

where K_n and K_s are joint normal and shear stiffnesses, σ_n and τ are the normal and shear stresses acting on the joint, and u_n and u_s are the normal and shear displacements, respectively.

　　The uniaxial compression test with a horizontal joint is designed to estimate the joint normal stiffness (Kulatilake et al., 2016). Note that there is no standard procedure in the ASTM or in the International Society of Rock Mechanics (ISRM) for this performed test. The samples were prepared similar to those used for the uniaxial compression test. A horizontal saw-cut joint was created at the mid-level of the sample, making it into two pieces with a length to diameter ratio of 1.0 for each. A layer of masking tape was wrapped around the horizontal joint to prevent the two pieces moving along the joint in the horizontal direction. To obtain the joint normal stiffness of bedding planes and joints among the same material, and interfaces and joints of two different materials, two halves from the same rock type and from two different materials were prepared as shown in Figure 6.13. The samples were placed in the test apparatus, shown in Figure 6.6, and loaded until they failed.

Figure 6.13 Samples used for uniaxial joint compression tests

6.3.6   *Small-scale direct shear test*

The small-scale direct shear test is generally used to estimate the mechanical properties, joint shear stiffness, joint friction angle and joint cohesion for discontinuities such as joints and bedding planes. In this research, the ASTM standard D5607 (2008) was followed. Samples with saw-cut joints were used. Each sample consists of two equal halves of cylindrical parts with a horizontal saw-cut joint placed at the middle. The thickness of each half is one inch (25 mm). Each half of the sample was securely mounted in hydrostone with the joint surface appearing outside (see Figure 6.14). The testing machine shown in Figure 6.15 was used with the top half of the sample fixed and bottom half sliding along the joint. A calculated weight was applied to provide normal stress in the vertical direction (perpendicular to the shear plane). The shear load was applied along the shear plane while keeping the normal stress constant until it reaches the maximum value—the peak shear strength. The joint shear stiffness can be calculated by the slope of the shear stress versus shear displacement curve prior to the peak shear strength; the M-C criterion for joints was used to determine the joint friction angle.

6.4   Results and discussions

For Brazilian tests, the samples that failed along pre-existing joints or chipped at the edges were ruled out; only the samples failed along the loading direction with the failure initialized by the tension at the center of the disc were considered as valid (Figure 6.4). Test results are given in Table 6.2. The sample numbers are 3 for dacite, 5 for ML, and 4 for LM. It can be seen that dacite has the lowest average tensile strength

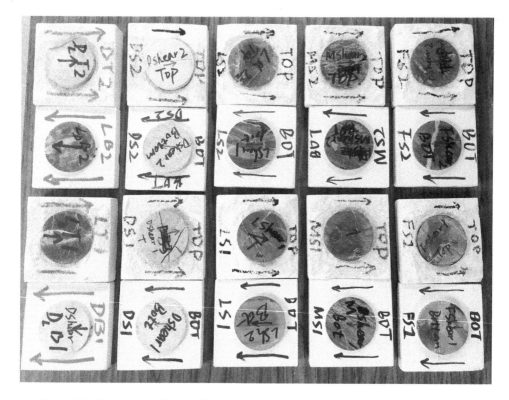

Figure 6.14 Samples used for small-scale direct shear tests

of 3.82 MPa, while the other two have higher values around 8.5 MPa. Due to the fact that the rock samples were taken from different depths and locations, as shown in Figure 6.1 and Table 6.1, moderate level deviations can be observed in the results for each rock type.

Based on the condition of original cores and the dimensional requirement of the test sample, the number of rock samples for the uniaxial compression test was less than that for the Brazilian test. After weighing and measuring the samples, the average density of each rock type was calculated and given in Table 6.3, which is 2380 kg/m³ for dacite and about 2740 kg/m³ for ML and LM. The UCS of the rocks is also presented in Table 6.3. A high deviation exists between the two dacite samples, which is likely caused by the different fracture distributions and failure modes. Both intact failure and shear failure along joints were observed for sample DU1, while a total joint shear failure took place in sample DU2. Moderate variations of the results for ML and LM rock samples can be noticed (Table 6.3). The factors such as different depths and locations of the rocks are mainly responsible for these deviations. Strain gauges were installed on five rock samples to determine the elastic constants, E and μ, and the results are given in Table 6.4. Dacite has the smallest Young's modulus and Poisson's ratio, 17.5 GPa and 0.2, respectively; ML and LM have close results of 70 GPa and 0.25, respectively. In fact, these values need more verification due to the small number of tests. The ultrasonic

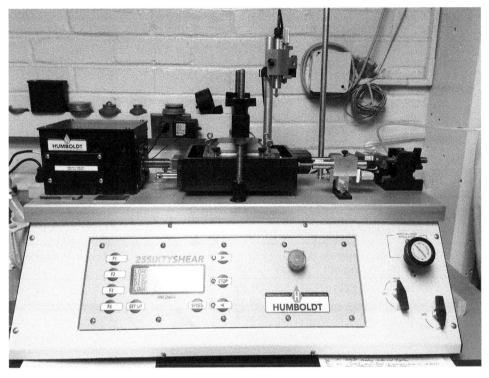

Figure 6.15 Testing machine used for small-scale direct shear tests

Source: Reprinted by permission from Springer Nature Customer Service Centre GmbH: Springer, *Geotechnical and Geological Engineering*, "Laboratory Estimation of Rock Joint Stiffness and Frictional Parameters," P. H.S.W. Kulatilake, S. Shreedharan, T. Sherizadeh, B. Shu, Y. Xing & P. He, 2017.

tests were hence conducted and are described in the next sub-section. Figure 6.16 shows the stress-strain curve for the dacite sample DU1. Similar plots of the other samples are shown in Figures A.1 through A.4 in Appendix A. Since the axial and lateral strain in these curves were obtained by the strain gauges, which were placed at the middle of the sample, the nonlinear part at low-stress level caused by pores and voids can hardly be observed.

Table 6.5 provides the ultrasonic test results for the rock samples. The densities of the samples are close to the values given in Table 6.3. The Young's modulus values are slightly higher than that estimated by the strain gauges (Table 6.4), and the results for each rock type are consistent. For Poisson's ratio, the results of ML and LM show a small variation, and close to the static values (see Table 6.4). An exception was obtained for dacite with an average value of 0.09. Inspection of the dacite samples showed many fractures and joints, which might have affected the traveling velocities of the waves.

Triaxial tests were carried out to estimate the strength parameters of the intact rock. A series of the maximum axial stress σ_1 (major principal stress) versus confining stress σ_3 (minor principal stress) were obtained for each rock type. Linear relations were assumed

Table 6.2 Brazilian test results

Rock type	Sample No.	Tensile strength (MPa)	Mean value (MPa)	Coefficient of variation
Dacite	DB1	2.85	3.82	0.40
	DB2	3.02		
	DB3	5.59		
ML	MB1	7.47	8.53	0.19
	MB2	10.01		
	MB3	6.18		
	MB4	9.63		
	MB5	9.34		
LM	LB1	9.31	8.51	0.41
	LB2	5.53		
	LB3	6.04		
	LB4	13.09		

Table 6.3 Results of uniaxial compressive strength of the samples

Rock type	Sample No.	Average density (kg/m^3)	UCS (MPa)	Mean value (MPa)	Coefficient of variation
Dacite	DU1	2380	75.22	49.7	0.72
	DU2		24.26		
ML	MU1	2746	116.63	136.7	0.22
	MU2		122.81		
	MU3		170.61		
LM	LU1	2740	192.73	151.4	0.29
	LU2		106.43		
	LU3		122.61		
	LU4		183.98		

Table 6.4 Results of elastic constants from uniaxial compression tests

Rock type	Sample No.	Young's modulus, E (GPa)	Poisson's ratio, μ	Mean value	
				E (GPa)	μ
Dacite	DU1	17.46	0.20	17.5	0.20
ML	MU1	62.95	0.26	73.0	0.25
	MU3	83.01	0.24		
LM	LU1	59.03	0.27	68.2	0.26
	LU4	77.27	0.24		

based on the Mohr-Coulomb failure criterion, which is given by Eqs. (6.4) and (6.5). After fitting the laboratory test results with the linear regression as shown in Figures 6.17 (the LM rock) and A.5 and A.6, the friction angle and cohesion were determined. The significant scatter in test results (Figures 6.17 and A.5 and A.6) is probably caused by the material heterogeneity, different locations and depths of the rock cores. Table 6.6 provides the triaxial test results for all the samples. The obtained UCS for dacite is 77.6 MPa, close to the value given by sample DU1 (75.22 MPa) in the uniaxial compression test (Table 6.3). The UCS of ML and LM are between 100 MPa and 130 MPa (Table 6.6), which is

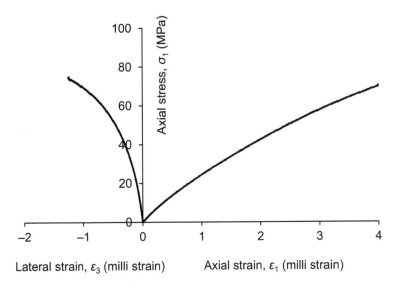

Figure 6.16 Stress-strain curve for DUI

Table 6.5 Ultrasonic test results

Rock type	Sample No.	Average Density (kg/m³)	E (GPa)	μ	Mean values		Coefficient of variation	
					E (GPa)	μ	E	μ
Dacite	DJKNIT	2430	26.30	0.06	25.0	0.09	0.08	0.48
	DJKN2B		23.60	0.12				
ML	MJKNIT	2749	70.10	0.27	73.4	0.25	0.06	0.08
	MJKNIB		78.00	0.23				
	MTI		72.00	0.27				
LM	LJKNIT	2742	65.30	0.22	68.7	0.25	0.05	0.10
	LJKN2T		68.90	0.26				
	LT3		71.80	0.26				

smaller than that in Table 6.3. The estimated friction angle and cohesion for all rock types are within the ranges of 30–40 degrees and of 22–37 MPa, respectively. The values of dacite are slightly lower than that of ML and LM.

$$\sigma_1 = \text{UCS} + \tan^2\left(45 + \frac{\phi}{2}\right) \cdot \sigma_3 \tag{6.4}$$

$$\text{UCS} = 2 \cdot c \cdot \tan\left(45 + \frac{\phi}{2}\right) \tag{6.5}$$

where σ_1 is the maximum principal stress (axial stress), σ_3 is the minimum principal stress (confining stress), ϕ is the friction angle, and c is cohesion.

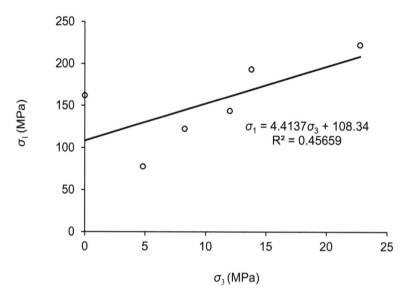

$$\sigma_1 = 4.4137\sigma_3 + 108.34$$
$$R^2 = 0.45659$$

Figure 6.17 Linear regression of triaxial test results for LM

Table 6.6 Triaxial test results

Rock type	Sample No.	σ_3 (MPa)	σ_1 (MPa)	UCS (MPa)	ϕ (°)	c (MPa)
Dacite	DTI	7.6	132.7	77.6	30.0	22.4
	DT2	15.2	142.1			
	DT3	22.8	122.5			
ML	MTI	4.0	167.0	127.8	29.7	37.1
	MT2	8.3	106.2			
	MT3	10.3	121.7			
	MT4	13.8	185.3			
	MT5	17.9	228.8			
	MT6	16.6	150.7			
LM	LTI	4.8	77.6	108.3	39.1	25.8
	LT2	8.3	122.5			
	LT3	12.0	143.9			
	LT4	13.8	193.4			
	LT5	22.8	222.5			

In the uniaxial joint compression test, the axial stress and displacement for each sample were recorded and plotted. This displacement is the total deformation of intact rock and joint. Taking sample DJKN1 as an example, a plot of normal stress versus total deformation was drawn (solid curve in Figure 6.18), which is nonlinear initially, but becomes linear after a certain point. Intuitively, by drawing a line passing through the origin point and parallel to the straight-line portion of the total deformation-normal stress curve, the deformation-stress curve of intact rock is obtained (see dashed line in Figure 6.18). Figure 6.19 shows the plot of joint deformation (D_j) versus normal stress (σ_n) for sample DJKN1 by subtracting intact deformation from the total deformation. A fitted exponential regression curve for the experimental results is shown in Figure 6.20. After getting the regression equation ($\sigma_n = 0.202e^{18.98D_j}$), the joint normal stiffness, K_n, can be estimated using the mathematical procedures given in Kulatilake et al. (2016):

$$\sigma_n = 0.202e^{18.987D_j} \tag{6.6}$$

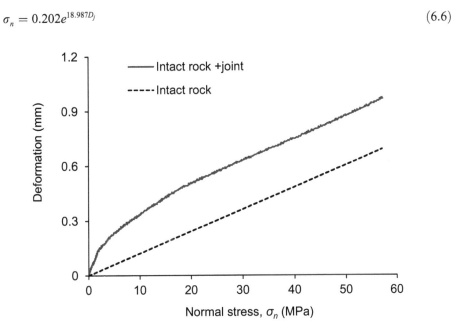

Figure 6.18 Diagram of total deformation versus normal stress and intact rock deformation versus normal stress for DJKN1

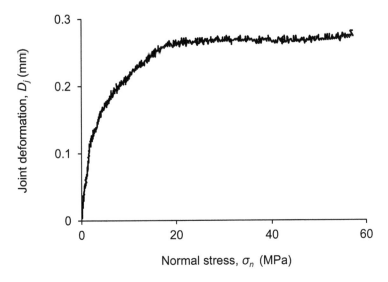

Figure 6.19 Diagram of joint deformation versus normal stress for DJKN1

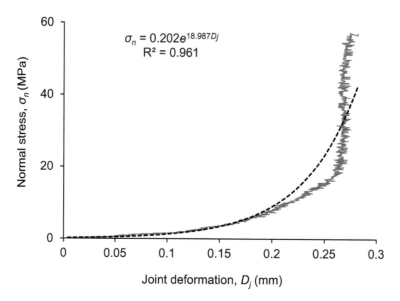

Figure 6.20 Diagram of normal stress versus joint deformation and the fitted exponential curve for dacite joint (DJKNI)

$$\ln \sigma_n = \ln 0.202 + 18.987 D_j \tag{6.7}$$

$$D_j = \frac{\ln \sigma_n - \ln 0.202}{18.987} \tag{6.8}$$

$$\frac{dD_j}{d\sigma_n} = \frac{1}{18.987\sigma_n} \tag{6.9}$$

$$K_n = \frac{d\sigma_n}{dD_j} = 18.987\sigma_n \tag{6.10}$$

Linear correlations were obtained between the joint normal stiffness and normal stress as shown in Eq. (6.10), and the coefficients (K_n/σ_n) of all the tests are summarized in Table 6.7. Under the same normal stress, the dacite joint has a higher K_n than the ML and LM joints. A moderate variation of the results can be seen in LM joints. The stiffness coefficient of the interface between dacite and LM gives the value (17.19×10^3/m), which seems within a reasonable range with respect to the values of the two rocks (20.68×10^3/m and 12.71×10^3/m). Similar to Figures 6.18 through 6.20, the diagrams that used to estimate the exponential functions for other rock samples are given in Figures A.7 through A.30.

In the small-scale direct shear tests, four normal loads were applied to each sample. Figure 6.21 shows the shear stress versus shear displacement curves at different normal stress levels for sample DS-1. By taking the slope of the curves, the joint shear stiffness (K_s) was determined. A linear relation was found between K_s and normal stress (σ_n) (Figure 6.22). After obtaining the plot of normal stress versus peak shear strength

Table 6.7 Results of uniaxial joint compression tests

Rock type	Sample No.	K_n/σ_n ($\times 10^3$ /m)	Mean values ($\times 10^3$/m)	Coefficient of variation
Dacite	DJKN1	18.99	20.68	0.12
	DJKN2	22.37		
ML	MJKN1	17.58	14.58	0.18
	MJKN2	12.70		
	MJKN3	13.47		
LM	LJKN1	15.48	12.71	0.31
	LJKN2	14.39		
	LJKN3	8.28		
Dacite, LM	DLJKN	17.19	17.19	-

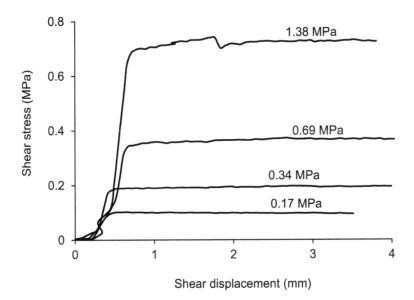

Figure 6.21 Diagram of shear stress versus shear displacement curves for dacite joint (DS-1)

(Figure 6.23), the joint friction angle and joint cohesion were estimated by using the Mohr-Coulomb criterion for joints (Eq. (6.11)).

$$\tau^p = \sigma_n \tan \phi_j + c_j \tag{6.11}$$

where τ^p is joint shear strength, σ_n is joint normal stress, ϕ_j is joint friction angle and c_j is joint cohesion.

For smooth joints, the joint cohesion (c_j) is nearly zero. Tables 6.8 and 6.9 summarize all the test results. The coefficient of joint shear stiffness over normal stress (K_s/σ_n) is in the range of 1.5–2.0; it is one-seventh to one-fifteenth of the value of K_n/σ_n (12–25 $\times 10^3$/m). The ML and LM joints have lower shear stiffness coefficients than the dacite joints. The results of ML show higher variations than that of others, which may be due to the man-made errors during sample preparation, such as slightly inclined joint surface and the uneven edge of the samples, or by the internal factors of samples, such as different joint

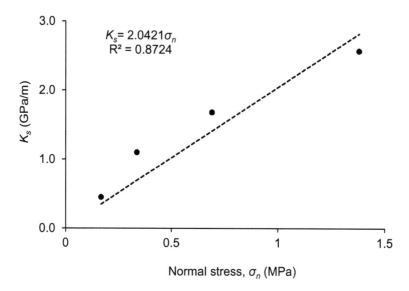

Figure 6.22 Diagram of K_s versus normal stress and the fitted regression curve for dacite joint (DS-1)

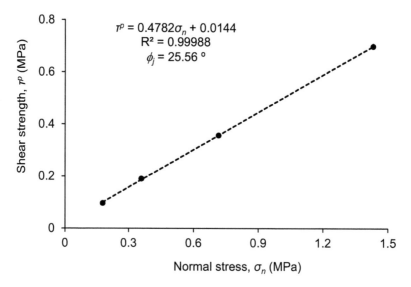

Figure 6.23 Diagram of shear strength versus normal stress and the fitted regression curve for dacite joint (DS-1)

patterns and different grain compositions. The interface between the dacite and LM provides reasonable shear stiffness values. With respect to the joint friction angle listed in Table 6.9, the average value of dacite joint is 28.6 degrees and about 30 degrees for the other two rock joints; small deviations of the results are observed for dacite, ML, and LM joints, but a relatively high deviation for the interface. Besides the aforementioned

Table 6.8 Summary of joint shear stiffness for all samples

Rock type	Sample No.	K_s(GPa/m) at different normal stresses (σ_n expressed in MPa)				K_s/σ_n ($\times 10^3$/m)	Mean value ($\times 10^3$/m)	Coefficient of variation
		$\sigma_n =$ 0.17	$\sigma_n =$ 0.34	$\sigma_n =$ 0.69	$\sigma_n =$ 1.38			
Dacite	DS-1	0.45	1.10	1.68	2.57	2.04	2.02	0.03
	DS-2	0.60	1.18	1.74	2.42	2.00		
ML	MS-1	0.61	0.89	1.36	1.93	1.59	1.50	0.37
	MS-2	0.37	0.69	1.06	1.83	1.38		
	MS-3	0.4	0.93	1.41	2.01	1.75		
	MS-4	0.38	0.75	1.4	2.27	1.87		
	MS-5	0.23	0.44	0.85	1.01	0.92		
LM	LS-1	0.32	0.56	1.18	2.34	1.70	1.72	0.16
	LS-2	0.59	0.96	1.53	2.29	1.84		
	LS-3	0.23	1.08	1.33	2.01	1.87		
	LS-4	0.82	0.94	1.17	2.28	1.47		
	LS-5	0.45	1.1	1.13	1.63	1.74		
Dacite, LM	LDS-1	0.37	0.66	1.54	1.99	1.62	1.58	0.04
	DLS-2	0.19	0.43	1.03	2.18	1.54		

Table 6.9 Summary of joint friction angle for all the samples

Rock type	Sample No.	ϕ_j (°)	Mean value (°)	Coefficient of variation
Dacite	DS-1	25.56	28.6	0.15
	DS-2	31.54		
ML	MS-1	27.01	31.4	0.13
	MS-2	27.73		
	MS-3	32.28		
	MS-4	33.76		
	MS-5	36.14		
LM	LS-1	39.30	30.1	0.19
	LS-2	27.30		
	LS-3	31.29		
	LS-4	25.99		
	LS-5	26.48		
Dacite, LM	LDS-1	29.75	29.75	0.49
	DLS-2	14.35		

reasons, the breakage of hydrostone cast of some samples (i.e. DS-1, MS-1, and DLS-2), as shown in Figure 6.24, could have impacted the test results. Figure 6.25 shows the edge failure of some samples. The rest of the plots of the test results is given in Figures A.31 through A.69.

6.5 Summary

Laboratory tests were carried out to estimate the mechanical properties of intact rock and smooth joints. Three types of rocks were tested and the results were presented and analyzed. The dacite is the weakest rock with the lowest tensile and shear strength, and the smallest

Figure 6.24 Broken hydrostone casts during tests

Figure 6.25 Failure at the edges of the direct shear samples

Young's modulus. Due to the pre-existing fractures and joints inside test samples, the dynamic Poisson's ratio of dacite is difficult to estimate. In addition, the dacite joint has higher stiffness coefficients (K_n/σ_n and K_s/σ_n) but lower basic friction angle (ϕ_j) than ML and LM joints. The ML and LM rocks have close values in densities, elastic constants and strengths, and similar results for joint properties. The joint stiffness and joint friction angle of the interface between dacite and LM are in reasonable ranges. Variations can be observed in the test results. Although the ASTM international standards were followed to prepare the samples and to execute the experiments, some operating deviations are unavoidable. Most importantly, since the rock samples were collected from boreholes at various depths and orientations, they are highly different in the number and distributions of fractures, grain compositions and so forth, causing different failure patterns of the samples.

Chapter 7

Three-dimensional discontinuum modeling of tunnel stability

7.1 Introduction

In this chapter, the tunnel stability of the area inside the square in Figure 2.2a is numerically simulated. Laboratory results of intact rock and rock joints discussed in Chapter 6 are used to estimate the rock mass properties and fault properties. Section 7.2 presents the estimation procedures of the rock mass properties which implicitly take into account the effect of minor discontinuities. The available geological, geotechnical and mine construction information provided by the mining company was used to build a three-dimensional discontinuum model, as given in Section 7.3. The model includes a weak inclined dike layer, the persistent and non-persistent faults, a complex tunnel system and the backfilling activities. The strain-softening block model and the continuously yielding joint model are respectively specified for the rock masses and the faults in the numerical model. Sequential constructions are implemented and delayed installation of rock supports is invoked. Sections 7.4 and 7.5 describe the performed analysis cases and the numerical results. First, preliminary results on model behavior and the impacts of complex geologies and engineering activities on tunnel stability are presented. Then parametric studies are carried out on the horizontal in-situ stress ratio, the rock mass conditions, the rock mass post-peak strengths, the fault properties and the delayed time of support installation. The overall tunnel stability is assessed using the distributions of stresses, displacements and failed zones around excavations, and the shear behavior of the major fault. The effectiveness of the rock support system is evaluated based on the rock mass behavior and the safety of supports. Field measurements from tape extensometers are used to make comparisons with numerical predictions. Finally, discussions are provided on the design of rock supports for this underground mine.

7.2 Estimation of the rock mass properties

As introduced in Chapter 2, the selected study region mainly consists of two lithologies, OC5 and Dike. Lithology OC5 is formed of limestone with thinly laminated to thinly bedded mudstones. Dike is an intrusive dacite layer with dip angles varying in the range of 25–45 degrees. The rock masses in OC5 have a wide range of RMR values, from 25 to 55 (poor to fair); the rock masses belonging to the dike is in a relatively poor condition with RMR values in the range of 25 to 40. The average RMR values for the two lithologies are respectively 40 and 32.5. The properties of intact rock and minor discontinuities are combined and represented as rock mass properties. The rock mass properties were estimated using the empirical formulas proposed by Hoek et al. (2002) and Hoek and Diederichs (2006).

According to Hoek and Brown (1997), the GSI (Geological Strength Index) (Hoek, 1994) can be estimated from the 1976 version of the RMR values (Bieniawski, 1976), with the groundwater rating set to 10 (dry) and the adjustment for joint orientation set to 10 (very favorable). Laboratory testing results are available for the intact rock. Due to the fact that both the LM and ML rocks contain limestone and mudstone, and they have similar mechanical property values (see the results in Chapter 6), the two rocks are considered as the intact rock of OC5, and their average properties are used.

The deformation modulus of the rock mass, E_{rm}, was determined by the following equation (Hoek and Diederichs, 2006):

$$E_{rm} = E_i \left(0.02 + \frac{1 - D/2}{1 + e^{((60 + 15D - GSI)/11)}} \right)$$

(7.1)

where E_i is Young's modulus of the intact rock; D is a factor describing the disturbance degree of the rock mass subjected by blast damage and stress relaxation. It varies from 0 for undisturbed rock masses to 1 for very disturbed rock masses. In this research, D was set to 0.1, representing the situation of low disturbance from blasting.

The rock mass strength parameters in the Hoek-Brown criterion are respectively given by

$$m_b = m_i \exp\left(\frac{GSI - 100}{28 - 14D}\right)$$

(7.2)

$$s = \exp\left(\frac{GSI - 100}{9 - 3D}\right)$$

(7.3)

$$a = \frac{1}{2} + \frac{1}{6}(e^{-GSI/15} - e^{-20/3})$$

(7.4)

where m_i is the intact rock parameter and can be obtained from the triaxial tests (Hoek and Brown, 1997).

By fitting the Hoek-Brown criterion with the Mohr-Coulomb envelope in a range of stress values, the Mohr-Coulomb parameters can be calculated by Eqs. (7.5) and (7.6) (Hoek et al., 2002).

$$\phi_{rm} = \sin^{-1}\left[\frac{6am_b(s + m_b\sigma_{3n})^{a-1}}{2(1 + a)(2 + a) + 6am_b(s + m_b\sigma_{3n})^{a-1}}\right]$$

(7.5)

$$c_{rm} = \frac{\sigma_{ci}[(1 + 2a)s + (1 - a)m_b\sigma_{3n}](s + m_b\sigma_{3n})^{a-1}}{(1 + a)(1 + 2a)\sqrt{1 + (6am_b(s + m_b\sigma_{3n})^{a-1})/((1 + a)(1 + 2a))}}$$

(7.6)

where ϕ_{rm} and c_{rm} are the friction angle and cohesion of the rock mass, σ_{ci} is the uniaxial compressive strength of the intact rock and σ_{3n} is a factor related to the maximum confining stress, σ_{3max}, within which the fitting relation between the Hoek-Brown and the Mohr-Coulomb criterion is considered. Hoek et al. (2002) provided guidelines to estimate the value of σ_{3max} for tunnels. According to that, the range of the confining stress used to obtain c_{rm} and ϕ_{rm} in this study is 0–8 MPa.

There are no available empirical equations suggested for the estimation of rock mass Poisson's ratio. However, Kulatilake et al. (2004) found that the Poisson's ratio increased about 21% from intact rock for rock masses due to the existence of discontinuities. The Poisson's

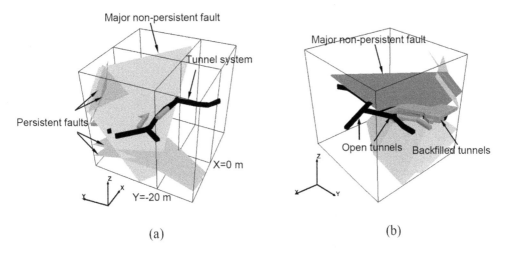

(a)   (b)

Figure 7.2 Faults and tunnel system built in the numerical model (represented by the outline box) seeing from views of (a) upper southwest and (b) upper northeast

Source: Reprinted from *Engineering Geology*, 238, Y. Xing, P.H.S.W. Kulatilake & L. A. Sandbak, "Effect of rock mass and discontinuity mechanical properties and delayed rock supporting on tunnel stability in an underground mine," pp. 62–75, 2018b, with permission from Elsevier.

Table 7.3 Strain-softening property values used for the rock masses in the numerical model[a]

Post-failure parameter	OC5	Dike
Residual friction angle, ϕ_r (°)	28.4 (80% of the peak[b])	21.8 (85% of the peak[c])
Residual cohesion, c_r (MPa)	0.78 (40% of the peak[b])	0.60 (50% of the peak[c])
Residual tensile strength, σ_{tr} (MPa)	1.02 (40% of the peak[b])	0.58 (50% of the peak[c])
e^{ps} $(e^{pt})^b$ (milli strain)	3	5

[a] Data from Y. Xing, P.H.S.W. Kulatilake & L. A. Sandbak, "Effect of rock mass and discontinuity mechanical properties and delayed rock supporting on tunnel stability in an underground mine," *Engineering Geology*, 238 (2018b): 62–75.
[b] e^{ps}, e^{pt} are the plastic strain parameters defined by Eqs. (5.7) and (5.8).
[c] Peak represents the strength values given in Table 7.2.

provides the post-failure property values for the rock masses, which were estimated by reviewing some references (Hajiabdolmajid et al., 2002; Ray, 2009; Alejano et al., 2012).

Laboratory results of smooth joints showed that the joint normal (K_n) and shear (K_s) stiffnesses are linearly related to the normal stress (σ_n). The continuously yielding joint model was hence applied for the faults to describe this kind of behavior. Due to the influence of filling material and aperture, the faults in the field could be significantly weaker than the smooth joints in the laboratory. The property values used for the faults in the model were taken as half of the property values obtained for the smooth joints (Tables 6.7 through 6.9) and are given in Table 7.4. Cohesion and tensile strength of faults were assumed as zero to be on the conservative side.

The presence of the non-persistent fault and the non-planar lithology interfaces requires the addition of fictitious joints (Kulatilake et al., 1992) to discretize the numerical model domain into polyhedral prior to performing stress analyses using 3DEC version 4.1 code. As far as the mechanical behavior is concerned, the fictitious joints should behave like the respective rock

Table 7.4 Mechanical property values used for discontinuities in the numerical model[a]

Joint type	Normal stiffness, K_n (GPa/m)	Shear stiffness, K_s (GPa/m)	Friction angle, ϕ_j (°)	Cohesion, c_j (MPa)	Tensile strength (MPa)
Fault in OC5	$6.82*\sigma_n{}^b$	$0.81*\sigma_n{}^b$	14	0	0
Fault in Dike	$10.34*\sigma_n{}^b$	$1.01*\sigma_n{}^b$	15	0	0
Discontinuity interfaces between OC5 and Dike	673.8	269.5	30.6	1.58	1.86
Fictitious joint in OC5	1169.5	467.8	35.5	1.96	2.56
Fictitious joint in Dike	178.2	71.3	25.6	1.20	1.15

[a] Data from Y. Xing, P.H.S.W. Kulatilake & L. A. Sandbak, "Effect of rock mass and discontinuity mechanical properties and delayed rock supporting on tunnel stability in an underground mine," *Engineering Geology*, 238 (2018b): 62–75.

[b] σ_n is the normal stress in MPa.

Table 7.5 Material property values used for rock supports in the numerical model[a]

Material property	Split sets	Resin bolts	Cable bolts	Swellex bolts
Young's modulus of bolt (GPa)	200	200	100	200
Bolt diameter (mm)	39	22	12.7	45
Cross-sectional area of bolt (m²)	1.19e−3	3.8e−4	1.27e−4	1.59e−3
Tensile yield capacity of the bolt (KN)	140	230	188	220
Bond shear stiffness (MN/m/m)	100	1350	5440	200
Bond strength (KN/m)	50	200	800	300

[a] Data from Y. Xing, P.H.S.W. Kulatilake & L. A. Sandbak, "Effect of rock mass and discontinuity mechanical properties and delayed rock supporting on tunnel stability in an underground mine," *Engineering Geology*, 238 (2018b): 62–75.

masses of the two lithologies with respect to both the strength and deformability. For fictitious joints and the discontinuity interfaces between different lithologies, the coulomb slip joint model was prescribed, where the mechanical property values were estimated using the method suggested by Kulatilake et al. (1992). In this research, the value of 0.008 was assigned to G/K_s, and 2.5 to the ratio K_n/K_s; the same strength parameter values were used for the rock masses and the fictitious joint. The discontinuity interfaces were assumed as well bonded, of which the mechanical property values were obtained by first calculating the average values between the two lithologies and then using the aforementioned method suggested by Kulatilake et al. (1992). Table 7.4 provides the mechanical property values of all the included discontinuities.

The three rounds of support installation described in Section 2.5.1 are numerically simulated for all the open tunnels in the numerical model. The rock bolts were simulated using the cable structure element, and the material property values used for these supports are given in Table 7.5. They were estimated based on the information provided by the mining company and by the manufacturers, and the suggestions given in the manual of 3DEC (see Section 5.6). The geometry and spacing of the rock bolt supports for the rectangular tunnels are similar to that applied for the horseshoe tunnels (see Section 2.5.1).

In the numerical model, the full-face tunneling method was assumed. According to the excavations conducted in the field, the open tunnels were excavated from 2004 to 2010, which was approximately divided into nine steps in the numerical model. Figure 7.3 presents

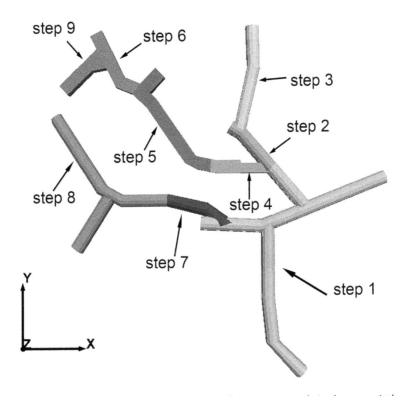

Figure 7.3 Plan view of the excavation sequence of the open tunnels in the numerical model

Source: Reprinted by permission from Springer Nature Customer Service Centre GmbH: Springer, *Rock Mechanics and Rock Engineering*, "Investigation of rock mass stability investigation around the tunnels in an underground mine in USA using three-dimensional numerical modeling," Y. Xing, P.H.S.W. Kulatilake & L. A. Sandbak, 2018a.

the excavation sequences of the open tunnels. The backfilled tunnels were also numerically excavated and backfilled step-by-step according to the construction time. Theoretically, the physical time of the excavation or supporting in the field is difficult to simulate by the numerical modeling. Nevertheless, it is related to the accumulated deformations of the unsupported rock mass. In other words, the moment when the supports are installed corresponds to a certain amount of occurred deformation or of the stress relaxation. Similar approaches to that applied by Vardakos et al. (2007) and Shreedharan and Kulatilake (2016) were adopted to take care of the delayed installation of supports. Detailed procedures were developed by incorporating a FISH function to 3DEC and are described as follows:

1   After excavation, the excavation boundaries were applied with the equivalent interior boundary tractions exerted by the excavated part, simulating the balance of unexcavated situation;
2   Reduced the interior boundary tractions by multiplying with the parameter $1 - \lambda$, where λ is the stress relaxation factor, and then cycled the model to equilibrium;
3   Installed the supports and removed the interior boundary tractions;
4   Cycled the model to equilibrium.

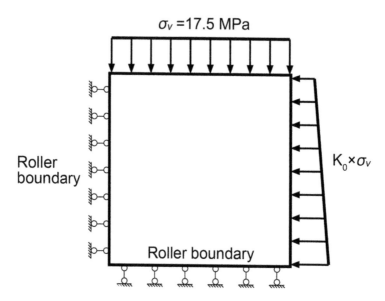

Figure 7.4 Boundary conditions applied for the numerical model

Source: Reprinted by permission from Springer Nature Customer Service Centre GmbH: Springer, *Rock Mechanics and Rock Engineering*, "Investigation of rock mass stability investigation around the tunnels in an underground mine in USA using three-dimensional numerical modeling," Y. Xing, P.H.S.W. Kulatilake & L. A. Sandbak, 2018a.

Procedures 1–4 were applied for each of the excavation-supporting steps. With respect to assuming a proper stress relaxation factor, the above two studies provided some advice. In this study, the factor was varied from zero to 80% at every 20% increment to assess the effects of delayed supporting on the rock mass behavior and on the safety of the supports.

Figure 7.4 shows the applied boundary conditions in the numerical model. Based on the depth of 647.7 m and overburden density of 2700 kg/m³, the vertical stress of 17.5 MPa was applied at the top boundary of the model to simulate overburden gravitational loading. The bottom boundary was specified with the roller boundary condition (no vertical velocity or displacement). Because of the asymmetric faults, lithologies and tunnels, stress boundaries cannot be specified at all the four side boundaries. Hence, the combined boundary conditions illustrated in Figure 7.4, rollers at one side and stresses at the other side are applied for both *x*- and *y*- directions. The vertical stress along the side boundary is increased according to the gravity gradient. The horizontal stresses are calculated from the horizontal in-situ stress ratio (K_0) × vertical stress (σ_v). For numerical modeling, K_0 was assigned with 0.5, 0.75, 1.0, 1.25, 1.5, and 2.0 (assuming the same K_0 value in *x*- and *y*-directions).

7.4   Performed stress analyses

Comprehensive investigations of the tunnel stability are conducted on the three-dimensional model. Detailed descriptions of the stress analysis cases are summarized in Table 7.6. Six K_0 values, three RMR systems (average, soft, and stiff), five residual strength levels, and three fault property conditions are included. The average values for each factor represent those given in the above sections and are explained in a footnote of Table 7.6. For the support

Table 7.6 Summary of the performed stress analyses

Case No.	K_0 value	RMR value	Residual strength	Fault property	Support system	Stress relaxation factor, λ
1(15)	0.5	Average[a]	Average[d]	Average[e]	U(S)	- (0[f])
2(16)	0.75	Average[a]	Average[d]	Average[e]	U(S)	- (0[f])
3(17)	1.0	Average[a]	Average[d]	Average[e]	U(S)	- (0[f])
4(18)	1.25	Average[a]	Average[d]	Average[e]	U(S)	- (0[f])
5(19)	1.5	Average[a]	Average[d]	Average[e]	U(S)	- (0[f])
6(20)	2.0	Average[a]	Average[d]	Average[e]	U(S)	- (0[f])
7(21)	1.0	Soft[b]	Average[d]	Average[e]	U(S)	- (0[f])
8(22)	1.0	Stiff[c]	Average[d]	Average[e]	U(S)	- (0[f])
9(23)	1.0	Average[a]	OC5: $c_r/\sigma_{tr} =$ 0.6*peak; $\phi_r =$ 0.9*peak Dike: $c_r/\sigma_{tr} =$ 0.7*peak; $\phi_r =$ 0.95*peak	Average[e]	U(S)	- (0[f])
10(24)	1.0	Average[a]	OC5: $c_r/\sigma_{tr} =$ 0.5*peak; $\phi_r =$ 0.85*peak Dike: $c_r/\sigma_{tr} =$ 0.6*peak; $\phi_r =$ 0.9*peak	Average[e]	U(S)	- (0[f])
11(25)	1.0	Average[a]	OC5: $c_r/\sigma_{tr} =$ 0.3*peak; $\phi_r =$ 0.75*peak Dike: $c_r/\sigma_{tr} =$ 0.4*peak; $\phi_r =$ 0.8*peak	Average[e]	U(S)	- (0[f])
12(26)	1.0	Average[a]	OC5: $c_r/\sigma_{tr} =$ 0.2*peak; $\phi_r =$ 0.7*peak Dike: $c_r/\sigma_{tr} =$ 0.3*peak; $\phi_r =$ 0.75*peak	Average[e]	U(S)	- (0[f])
13(27)	1.0	Average[a]	Average[d]	Joint stiffness = 5*average stiffness; joint friction angle = 21°	U(S)	- (0[f])
14(28)	1.0	Average[a]	Average[d]	Joint stiffness = 0.1*average stiffness; joint friction angle = 7°)	U(S)	- (0[f])
29	1.0	Average[a]	Average[d]	Average[e]	S	20%
30	1.0	Average[a]	Average[d]	Average[e]	S	40%
31	1.0	Average[a]	Average[d]	Average[e]	S	60%
32	1.0	Average[a]	Average[d]	Average[e]	S	80%
33	1.0	Average[a]	Average[d]	Average[e]	S	50%

[a] RMR(OC5) = 40, RMR(Dike) = 32.5.
[b] RMR(OC5) = 36.5, RMR(Dike) = 29.
[c] RMR(OC5) = 43.5, RMR(Dike) = 36.
[d] Residual strength values given in Table 7.3.
[e] Fault property values given in Table 7.4.
[f] Instantaneous installation of the supports.

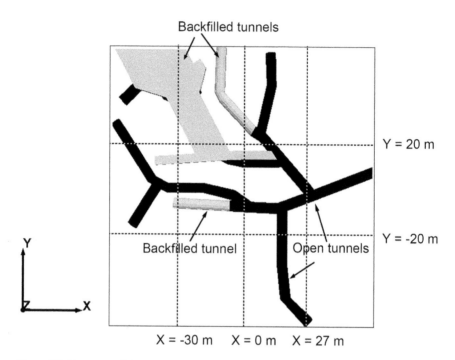

Figure 7.5 Plan view of the selected vertical planes to present the representative results
Source: From Xing et al. (2019).

system, U represents the unsupported cases; S indicates the supported cases applied with rock and cable bolts, which has been described in Section 2.5.1. Multiple stress relaxation factors are applied for the support system, as listed in Table 7.6, where the value of "0" (zero) represents the instantaneous supporting.

7.5   Numerical results and discussions

In this section, numerical modeling results are presented and discussed. Most of the results are obtained from the two representative vertical planes of X = 0 m and Y = −20 m, as shown in Figures 7.2a and 7.5. The distances between the tunnels on the planes and the model boundaries are more than eight times the tunnel dimensions; that means long enough to avoid the boundary effects. To account for the influence of complex geologies and engineering construction activities, several other cross-sectional planes (i.e. X = −30 m, X = 27 m, and Y = 20 m) are selected to show the results (see Figure 7.5).

7.5.1   Check of model behavior and the effect of complicated geologies and engineering activities

The model behavior is first checked using the results of Case 3 (see Table 7.6). Figures 7.6 and 7.7 show the stress distributions on the vertical planes of X = 0 m and Y = −20 m. Negative stress values represent compression. On the vertical plane of X = 0 m (Figure 7.6), two open tunnels exist; the major non-persistent fault goes in between the tunnels; the weak dike

ZZ Stress

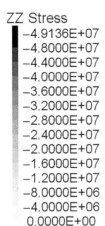

-4.9136E+07
-4.8000E+07
-4.4000E+07
-4.0000E+07
-3.6000E+07
-3.2000E+07
-2.8000E+07
-2.4000E+07
-2.0000E+07
-1.6000E+07
-1.2000E+07
-8.0000E+06
-4.0000E+06
0.0000E+00
1.4426E+06

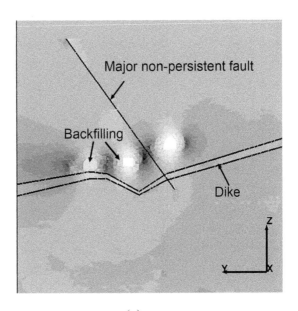

(a)

YY Stress

-3.3264E+07
-3.2500E+07
-3.0000E+07
-2.7500E+07
-2.5000E+07
-2.2500E+07
-2.0000E+07
-1.7500E+07
-1.5000E+07
-1.2500E+07
-1.0000E+07
-7.5000E+06
-5.0000E+06
-2.5000E+06
0.0000E+00
1.4426E+06

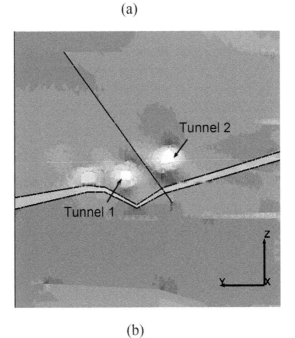

(b)

Figure 7.6 Stress distributions for Case 3 on the vertical plane of X = 0 m (unit: Pa): (a) vertical stress (ZZ stress) distribution; (b) horizontal stress (YY stress) distribution

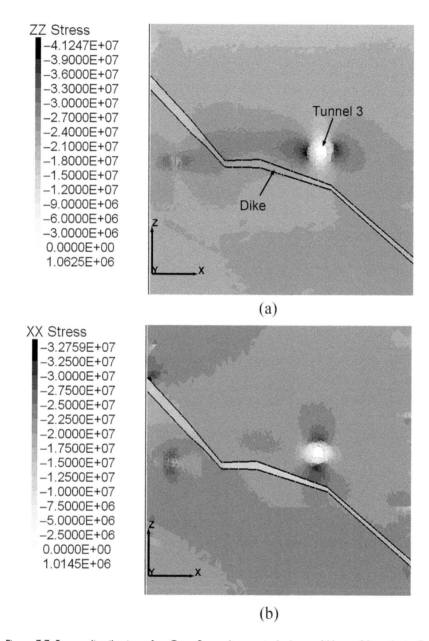

ZZ Stress
- -4.1247E+07
- -3.9000E+07
- -3.6000E+07
- -3.3000E+07
- -3.0000E+07
- -2.7000E+07
- -2.4000E+07
- -2.1000E+07
- -1.8000E+07
- -1.5000E+07
- -1.2000E+07
- -9.0000E+06
- -6.0000E+06
- -3.0000E+06
- 0.0000E+00
- 1.0625E+06

Tunnel 3

Dike

(a)

XX Stress
- -3.2759E+07
- -3.2500E+07
- -3.0000E+07
- -2.7500E+07
- -2.5000E+07
- -2.2500E+07
- -2.0000E+07
- -1.7500E+07
- -1.5000E+07
- -1.2500E+07
- -1.0000E+07
- -7.5000E+06
- -5.0000E+06
- -2.5000E+06
- 0.0000E+00
- 1.0145E+06

(b)

Figure 7.7 Stress distributions for Case 3 on the vertical plane of Y = −20 m (unit: Pa): (a) vertical stress (ZZ stress) distribution; (b) horizontal stress (XX stress) distribution

layer is located below the two tunnels; the backfilling activities occurred at the left of the left-side tunnel. The two tunnels are labeled as numbers 1 and 2 (see Figure 7.6b). As shown in Figure 7.6a, high vertical stresses (ZZ stress) are concentrated on the walls and peaks near the fault (values in dark colors); stress relaxation can be observed on the roof and the floor (values in white colors). The horizontal stresses (YY stress) (Figure 7.6b), however, are nearly zero on the ribs but high on the roof and the floor; the highest stresses tend to distribute near the dike and the fault. Low horizontal stresses can be seen in the dike layer. On the vertical plane of $Y = -20$ m (see Figure 7.7), there is only one tunnel (labeled as Tunnel 3) with the dike underneath. Under this condition, the stress distributions are more symmetric compared to that presented in Figure 7.6, and the maximum stress values are lower. Low horizontal stresses can also be observed in the dike layer in Figure 7.7b. All figures showed that far-field stresses have not been affected by the excavations. The vertical stress at the top boundary is around 18 MPa (see Figures 7.6a and 7.7a); the horizontal stresses at side boundaries ($Y = -61$ m and $X = 61$m) are in the range of 17.5–21 MPa (see Figures 7.6b and 7.7b), agreeing with the applied boundary stresses, 17.5 MPa at the top and 17.5–20.5 MPa on the side boundaries, where $K_0 = 1.0$ (Case 3).

Figure 7.8 show the vertical (Z-) and horizontal (Y-) displacement distributions on the vertical plane of $X = 0$ m. The rock masses in the footwall of the fault suffer high vertical displacements, and the variation pattern is parallel with the fault trace (Figure 7.8a). The maximum roof sag of 7.94 cm can be observed for Tunnel 1. In addition, the maximum horizontal displacement (+6.62 cm) occurs on the right rib of Tunnel 1 (Figure 7.8b), close to the major fault. These high values and distribution pattern of the displacement are caused by the nearby frequent excavation activities, backfilling and the existence of the fault. Figure 7.9 present the deformation distributions on the vertical plane of $Y = -20$ m. By contrast, the deformations around the tunnel are smaller due to the less disturbance. Slight asymmetric distribution of the vertical displacement above the tunnel is noticed (Figure 7.9a), probably due to the existence of the weak layer. Note that this is a three-dimensional model, and the plane of $Y = -20$ m is not isolated. Thus, the displacement distribution can also be affected by nearby out-of-plane activities.

Since the influence of complex geologies and of engineering activities have been observed in the above results, further analysis is presented on the multiple planes shown in Figure 7.5. Figure 7.10 shows the distributions of displacement vectors around open tunnels on the planes. On the plane of $X = -30$ m (see Figures 7.5 and 7.10a), one open tunnel exists with one backfilled tunnel on the left and the other above the right-side area (as labeled in Figure 7.10a); the weak dike layer is located on the roof. The maximum displacement of around 12.5 cm occurs on the roof of the tunnel. On the plane of $X = 0$ m (see Figures 7.5 and 7.10b), two open tunnels (Tunnels 1 and 2 in Figures 7.6b) are located respectively on the footwall and the hanging wall of the major non-persistent fault. Several locations to the left of the left-side tunnel were excavated and backfilled. Because of the many engineering activities on the footwall and the relief from the fault, the left-side tunnel suffers higher deformations than the right-side one, especially on the footwall and the right wall. The maximum displacement is around 8 cm on the roof. On the plane of $X = 27$ m (Figures 7.5 and 7.10c), a tunnel intersection exists below the major fault at a relatively far distance. Even though the tunnel intersection has a wider span, the deformations are smaller than the two previous conditions due to the fewer engineering disturbances and better geological conditions. On the plane of $Y = 20$ m (Figures 7.5 and 7.10d), large deformations, with the peak value of 12.0 cm, are observed above the left-side tunnel, where the weak dike layer, the fault, and the extensive excavation and backfilling activities present. Significant movements take place on the middle part of the right wall due to the existence of surrounding backfill and nearby weak dike layer. By comparison, the right-side tunnel is located far away from the complicated geological conditions and engineering activities, so that it is relatively stable. The displacements around the tunnel on the

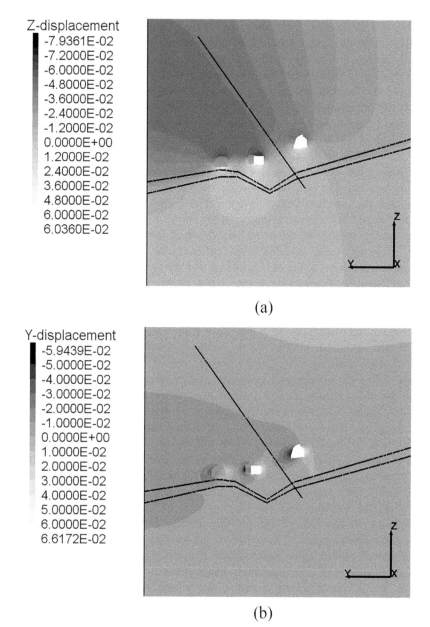

Figure 7.8 Distributions of the displacements around the tunnels for Case 3 on the vertical plane of X = 0 m (unit: m): (a) vertical (Z-) displacement; (b) horizontal (Y-) displacement

Source: Reprinted from *Engineering Geology*, 238, Y. Xing, P.H.S.W. Kulatilake & L. A. Sandbak, "Effect of rock mass and discontinuity mechanical properties and delayed rock supporting on tunnel stability in an underground mine," pp. 62–75, 2018b, with permission from Elsevier.

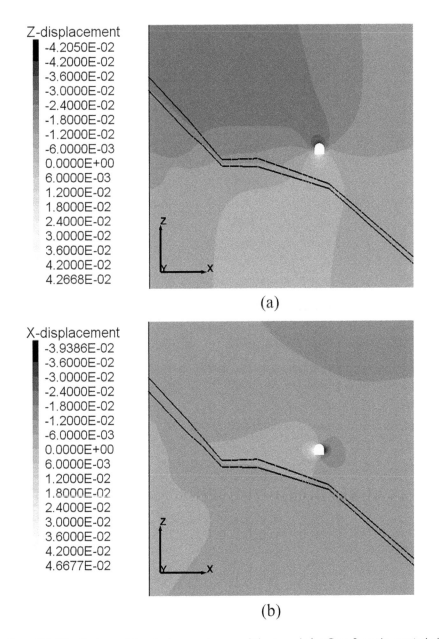

Z-displacement

-4.2050E-02
-4.2000E-02
-3.6000E-02
-3.0000E-02
-2.4000E-02
-1.8000E-02
-1.2000E-02
-6.0000E-03
0.0000E+00
6.0000E-03
1.2000E-02
1.8000E-02
2.4000E-02
3.0000E-02
3.6000E-02
4.2000E-02
4.2668E-02

(a)

X-displacement

-3.9386E-02
-3.6000E-02
-3.0000E-02
-2.4000E-02
-1.8000E-02
-1.2000E-02
-6.0000E-03
0.0000E+00
6.0000E-03
1.2000E-02
1.8000E-02
2.4000E-02
3.0000E-02
3.6000E-02
4.2000E-02
4.6677E-02

(b)

Figure 7.9 Distributions of the displacements around the tunnels for Case 3 on the vertical plane of Y = −20 m (unit: m): (a) vertical (Z-) displacement; (b) horizontal (X-) displacement

Source: Reprinted from *Engineering Geology*, 238, Y. Xing, P.H.S.W. Kulatilake & L. A. Sandbak, "Effect of rock mass and discontinuity mechanical properties and delayed rock supporting on tunnel stability in an underground mine," pp. 62–75, 2018b, with permission from Elsevier.

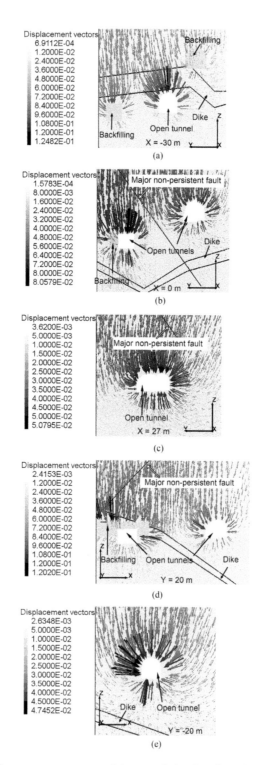

Figure 7.10 Displacement vectors around the tunnels for Case 3 on the vertical planes (unit: m): (a) X = −30 m; (b) X = 0 m; (c) X = 27 m; (d) Y = 20 m; (e) Y = −20 m

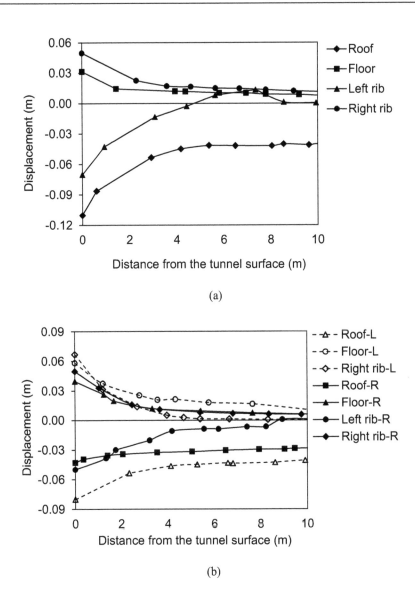

Figure 7.11 In-rock displacements around the tunnels for Case 3 on the vertical planes: (a) X = −30 m; (b) X = 0 m; (c) X = 27 m; (d) Y = 20 m; (e) Y = −20 m

plane of Y = −20 m seem to be the lowest, with the maximum value of 4.7 cm; the slight asymmetric displacement distribution could be caused by the dike layer that exists below.

To show the interior rock mass behavior starting from the excavation surfaces, vertical lines are placed above the roofs and below the floors to display the vertical displacements, and horizontal lines perpendicular to tunnel walls are used to provide the horizontal displacements. Figures 7.11(a–e) plot the variation of displacements with the distance from the tunnel surface for the various tunnels. For the planes where two open tunnels exist (i.e. X = 0 m and Y = 20 m), the descriptions of "-L" and "-R" in the legend in Figures 7.11b and 7.11d

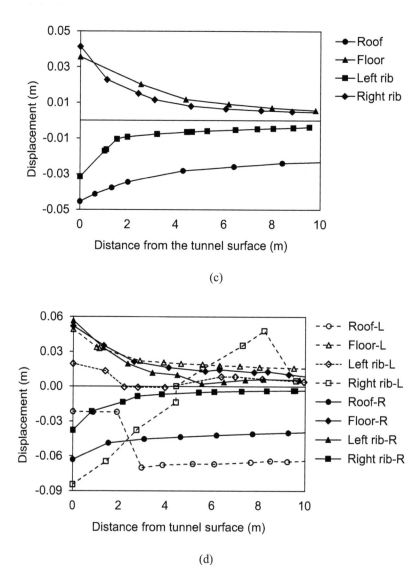

(c)

(d)

Figure 7.11 (Continued)

represent the left-side and right-side tunnels, respectively. Results show that large displacements are normally concentrated within the 2 or 3 m thickness from the tunnel surface; after 4 m, the movements either change slightly or stay constant with the distance. However, some deviations and distinctions can be found. Corresponding to the results presented in Figure 7.10a, Figure 7.11a illustrates that the tunnel on the plane of X = −30 m has a considerable roof sag of 11 cm, and a deep extent of high displacement on the roof, approximately 4 m. As shown in Figure 7.11b, the roof and the right wall of the left-side tunnel are pretty active, with displacements of 8.0 cm (Roof-L) and 6.7 cm (Right rib-L); they also have greater extents of high displacements than the other locations. Situations are more complicated on

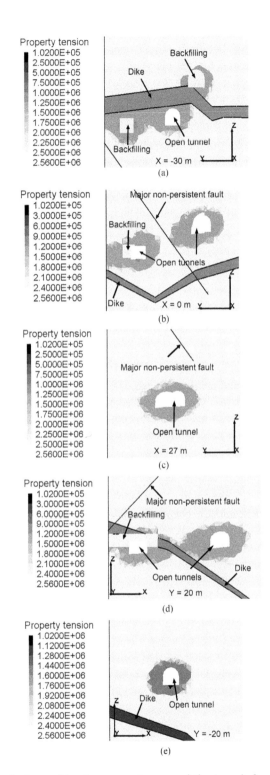

Figure 7.13 Distributions of tension parameter around the tunnels for Case 3 on the vertical plane (unit: Pa): (a) X = −30 m; (b) X = 0 m; (c) X = 27 m; (d) Y = 20 m; (e) Y = −20 m

of Y = 20 m (Figure 7.12d), extensive failures can be observed in-between the tunnels, resulting in unstable conditions. On the plane of Y = −20 m (Figure 7.12e), the failures are the least, and distribution is more or less symmetric.

In summary, the rock mass behavior for Case 3 has been presented through the distributions of the stress, displacement, and strength parameter values. The results were found to be reasonable intuitively, indicating that the numerical model responds correctly. On the other hand, the presence of complex geologies (i.e. the fault and the weaker layer) and the engineering disturbances have caused great influence on the stability of tunnels, resulting in large deformations and unstable rock masses to the nearby excavations. Results also showed that the rock masses within a distance of 2 or 3 m from the excavation surface present active movements, and those of 4–5 m away are stable with low displacements; exceptions can be observed at some locations with the unfavorable geologies and extensive excavation activities.

7.5.2   Effect of horizontal in-situ stress ratio (K_0) on tunnel stability

The effect of the horizontal in-situ stress on rock mass behavior is investigated using the results of Cases 1–6 (see Table 7.6). Figures 7.14a–f show the principal stress distributions for these cases on the plane of X = 0 m. It can be seen that the directions of principal stresses follow the shape of the tunnels and that the stress perpendicular to the tunnel surface is zero. For the low horizontal stress conditions (K_0 = 0.5, 0.75), the stress concentrations occur near the walls of the two tunnels and close to the geologies, fault and dike, as shown in Figures 7.14a and 7.14b. Maximum stress values appear adjacent to the left wall of Tunnel 1. The stresses on the roof and floor are small with a few positive values (tension). For Cases 3 and 4 (Figures 7.14c and 7.14d), large stress values have the tendency to move to the roofs and floors. As K_0 increases to 1.5 and 2.0 (Figures 7.14e and 7.14f), the high compressive stresses are appearing on the roofs and the floors, and at multiple locations near the fault and the dike. Stress relaxation in the dike layer is distinct for the last two cases. It is likely that the weak material is not able to sustain high stresses, which is hence transferred to the adjacent stronger rock masses. The maximum value of major principal stress reduces at first, from 54.5 MPa in Case 1 (K_0 = 0.5) to the lowest 53.1 MPa in Case 3 (K_0 = 1.0), but then increases to the peak of 77.6 MPa in Case 6 (K_0 = 2.0). Similar results can be found for the distribution of principal stresses on the plane of Y = −20 m (see Figures 7.15a–f). Because of the relatively favorable geologies, the distributions of principal stresses seem more symmetric, and the gradual change of stress concentration location is well presented, from tunnel walls at low K_0 cases to the roofs and floors at high K_0 cases. Slight asymmetric is observed due to the existence of the weak layer. The maximum major principal stress also minimizes at K_0 = 1.0 and peaks at K_0 = 2.0.

Plots in Figures 7.16 and 7.17 show the variations of the horizontal and vertical convergences of the tunnels, respectively. Different symbols represent the results of different tunnels on the two vertical planes. The displacements used to calculate the convergence were taken from the middle of the roof, ribs, and floor of the tunnels. Results show that most of the convergences, in both directions, increase slightly from K_0 = 0.5 to K_0 = 1.25 but distinctly from K_0 = 1.25 to K_0 = 2.0. For Tunnel 1, due to the fact that the left rib was backfilled, the value of horizontal convergence is less than that of the other two tunnels, as presented in Figure 7.16; however, it suffered more vertical convergence in cases K_0 = 0.5, 0.75, 1.0 and 1.25 (see Figure 7.17). For Tunnel 2 and Tunnel 3, the results show that at low K_0 values, 0.5

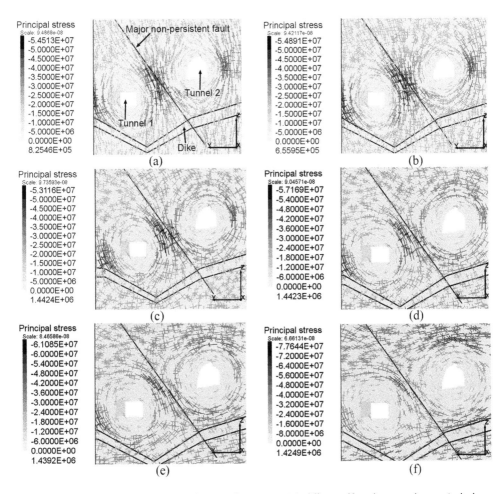

Figure 7.14 Principal stress distributions for cases with different K_0 values on the vertical plane of X = 0 m (unit: Pa). (a) Case 1; (b) Case 2; (c) Case 3; (d) Case 4; (e) Case 5; (f) Case 6

and 0.75, for example, the horizontal convergences are slightly higher than the vertical convergences; at high K_0 values (1.5 and 2.0), the vertical convergences are significant.

As previously mentioned, the rock masses that reach their residual strength, either in compression or in tension, are considered as failed. In addition, most of the failure around tunnels was found to be the shear failure. Therefore, the value distribution of the cohesion parameter is evaluated hereinafter. To quantify the failed area, the failure zone thicknesses were measured at three locations on each surface and the average values are provided in Figures 7.18 and 7.19. Since the failing rock masses on the left side of Tunnel 1 includes the backfilling area (see Figure 7.12b), measurements are absent at this location. Figure 7.18 shows that failure zone thicknesses on the ribs do not vary monotonically with K_0, but most of them slightly reduce from $K_0 = 0.5$ to $K_0 = 2.0$. However, Figure 7.19 shows a significant increase in the failure zone thicknesses on the roofs and floors of the tunnels when the K_0 value increases. These changes

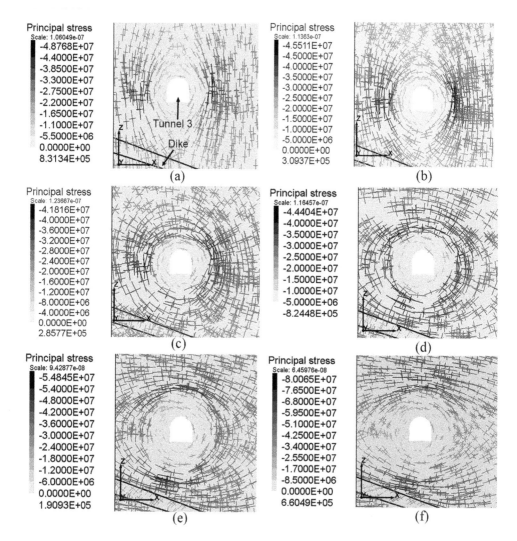

Figure 7.15 Principal stress distributions for cases with different K_0 values on the vertical plane of Y = −20 m (unit: Pa): (a) Case 1; (b) Case 2; (c) Case 3; (d) Case 4; (e) Case 5; (f) Case 6

seem to coincide with the results of the stress variations presented in Figures 7.14 and 7.15; the cases with high K_0 values have unfavorable roof and floor conditions but relatively stable ribs. Conversely, for low K_0 cases (i.e. 0.5 and 0.75), more failures are taking place on the ribs than on the roofs and floors (see Figures 7.18 and 7.19). The two locations, the right rib of Tunnel 1 and the left rib of Tunnel 2 (Figure 7.18), have higher values than other places. It might be ascribed for the close distance to the major fault (Figure 7.14a). For Tunnel 3, the rock masses on the floor are undergoing more failure than that on the roof (Figure 7.19), which is likely due to the influence of the dike located below the floor (Figure 7.15a).

Figures 7.20 shows the joint shear behavior of the major non-persistent fault with different K_0 values. The vectors represent the shear displacements of the hanging wall plane of the

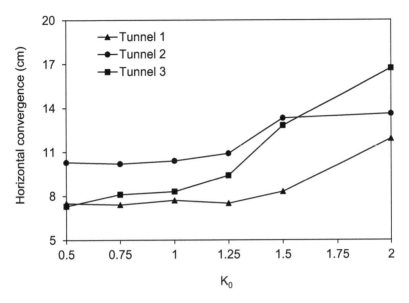

Figure 7.16 Horizontal convergence of the tunnels for cases with different K_0 values

Source: Reprinted by permission from Springer Nature Customer Service Centre GmbH: Springer, *Rock Mechanics and Rock Engineering*, "Investigation of rock mass stability investigation around the tunnels in an underground mine in USA using three-dimensional numerical modeling," Y. Xing, P.H.S.W. Kulatilake & L. A. Sandbak, 2018a.

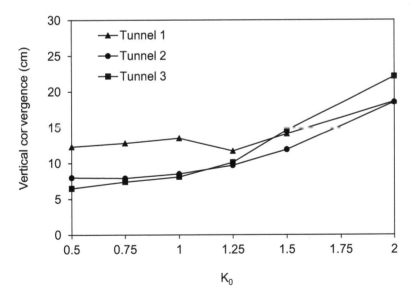

Figure 7.17 Vertical convergence of the tunnels for cases with different K_0 values

Source: Reprinted by permission from Springer Nature Customer Service Centre GmbH: Springer, *Rock Mechanics and Rock Engineering*, "Investigation of rock mass stability investigation around the tunnels in an underground mine in USA using three-dimensional numerical modeling," Y. Xing, P.H.S.W. Kulatilake & L. A. Sandbak, 2018a.

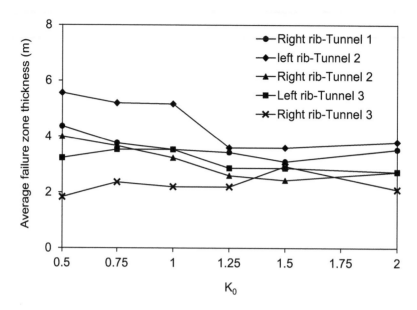

Figure 7.18 Average failure zone thicknesses on the ribs of the tunnels for cases with different K_0 values

Source: Reprinted by permission from Springer Nature Customer Service Centre GmbH: Springer, *Rock Mechanics and Rock Engineering*, "Investigation of rock mass stability investigation around the tunnels in an underground mine in USA using three-dimensional numerical modeling," Y. Xing, P.H.S.W. Kulatilake & L. A. Sandbak, 2018a.

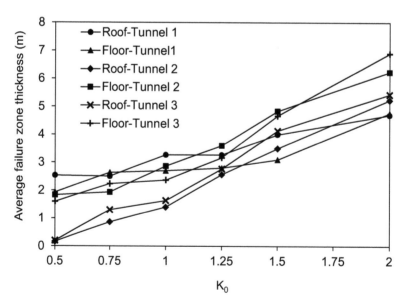

Figure 7.19 Average failure zone thicknesses on the roofs and floors of the tunnels for cases with different K_0 values

Source: Reprinted by permission from Springer Nature Customer Service Centre GmbH: Springer, *Rock Mechanics and Rock Engineering*, "Investigation of rock mass stability investigation around the tunnels in an underground mine in USA using three-dimensional numerical modeling," Y. Xing, P.H.S.W. Kulatilake & L. A. Sandbak, 2018a.

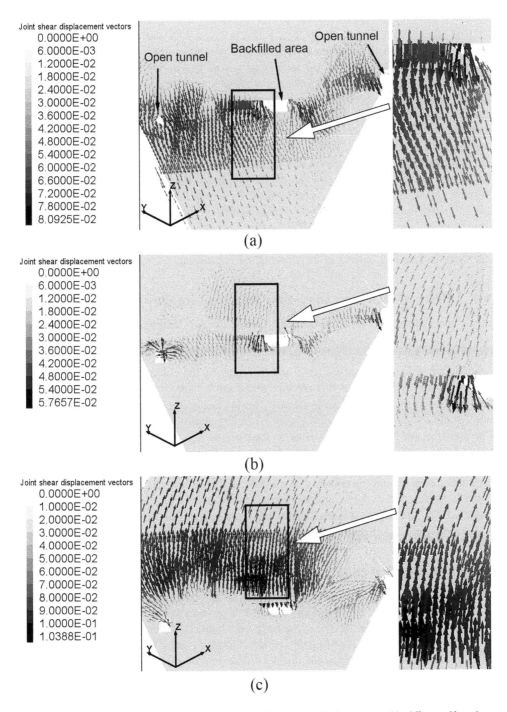

Figure 7.20 Joint shear displacement vectors along the fault for cases with different K_0 values (unit: m) (observed from the hanging wall direction): (a) Case 1; (b) Case 3; (c) Case 6

Source: Reprinted by permission from Springer Nature Customer Service Centre GmbH: Springer, *Rock Mechanics and Rock Engineering*, "Investigation of rock mass stability investigation around the tunnels in an underground mine in USA using three-dimensional numerical modeling," Y. Xing, P.H.S.W. Kulatilake & L. A. Sandbak, 2018a.

fault (see Figure 7.2a). To show the results clearly, a selected area on the fault plane is enlarged. Three representative cases were chosen to present results. For Case 1, where the K_0 value is 0.5 (see Figure 7.20a), the fault hanging wall moves downwards; displacements are mainly occurring at the lower half plane; the maximum value of 8.1 cm is located at the backfilled region. For Case 3 ($K_0 = 1.0$), the displacement values become much smaller; slight movements toward up can be observed at the upper part while the lower part is moving down; the maximum value of 5.8 cm is still near the backfilled region (see Figure 7.20b). As K_0 value increases to 2.0 (Figure 7.20c), the fault plane, however, shows distinct movements to up; the maximum displacement is 10.4 cm, above the backfilled area. The plots of joint shear displacement for other cases are given in Figure B.1. Figure 7.21 provides the maximum values of the fault shear displacement for Cases 1–6; Cases 3 ($K_0 = 1.0$) and 4 ($K_0 = 1.25$) give the lowest values. The K_0 value significantly changes the shear displacement of the major non-persistent fault in both the direction and magnitude. This change would result in totally different reactions on the rock mass in the numerical model, contributing to the deformations and failures of the surrounding excavations.

7.5.3 Effect of rock mass conditions on tunnel stability

To take into account the variation of RMR values, the soft, average and stiff rock mass systems (Cases 7, 3 and 8) are studied. Results of deformations and failure zones around tunnels can be found in Figures 7.22 and 7.23. The multiple locations are positioned on the horizontal axis; the numbers 1, 2 and 3 represent the different tunnels labeled in Figure 7.14a and Figure 7.15a. It is obvious that Case 7 (Soft) with the lowest rock mass properties results in the largest deformations (circles in Figure 7.22), approximately three times that of Case 8 (Stiff) (squares in Figure 7.22). Because of the more disturbances, backfilling activities and major faults, existing on the plane of $X = 0$ m, the deformations around Tunnels 1 and 2 are larger than that around Tunnel 3. This phenomenon is more distinct in the soft case (circles in Figure 7.22). Figure 7.23 shows that the soft system (Case 7) has the largest failed area around the tunnels as almost twice that of the stiff system (Case 8). High values can be observed at the two locations close to the major fault (right rib-1 and left rib-2 in Figure 7.23). The influence of rock mass properties on the shear displacement of the fault is shown in Figure B.2. Some effect is observed on the maximum displacement values.

7.5.4 Effect of post-failure parameters on tunnel stability

Results of Cases 3 and 9–12 (see Table 7.6) are used to investigate the influences of rock mass post-failure strengths on the tunnel stability. Figure 7.24 presents the variations of the maximum horizontal displacements on the ribs of the three open tunnels. As the residual strength gets smaller, the deformations on the ribs keep rising at increasing rates. For Cases 11 and 12, the deformation increments are significant, distinct on the right rib of Tunnel 1 and the left rib of Tunnel 2 (dashed line with circles and solid line with squares in Figure 7.24). Note that the two locations are close to the fault. Figure 7.25 shows the maximum vertical displacements on the roofs and the floors for the five cases. Similarly, the deformations increase with decreasing residual strength. The increasing trend is milder than that on the ribs. Relatively large values can be observed on the roof of Tunnel 1 (solid line with circles in Figure 7.25).

As shown in Figures 7.26 and 7.27, the area of the failed rock masses present distinct increases from Case 9 to Case 12. For Case 9, the average failure zone thicknesses are

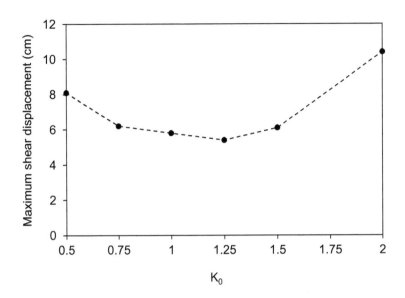

Figure 7.21 Maximum shear displacement on the fault plane for cases with different K_0 values

Source: Reprinted by permission from Springer Nature Customer Service Centre GmbH: Springer, *Rock Mechanics and Rock Engineering*, "Investigation of rock mass stability investigation around the tunnels in an underground mine in USA using three-dimensional numerical modeling," Y. Xing, P.H.S.W. Kulatilake & L. A. Sandbak, 2018a.

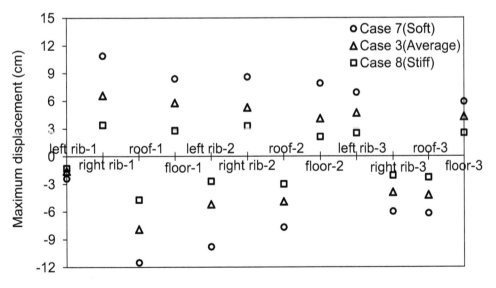

Figure 7.22 Maximum displacements around the tunnels for different rock mass systems

Source: Modified by permission from Springer Nature Customer Service Centre GmbH: Springer, *Rock Mechanics and Rock Engineering*, "Investigation of rock mass stability investigation around the tunnels in an underground mine in USA using three-dimensional numerical modeling," Y. Xing, P.H.S.W. Kulatilake & L. A. Sandbak, 2018a.

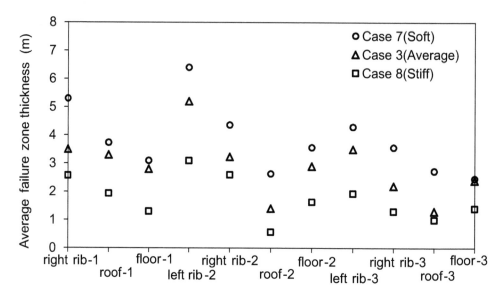

Figure 7.23 Average failure zone thicknesses around the tunnels for different rock mass systems

Source: Modified by permission from Springer Nature Customer Service Centre GmbH: Springer, *Rock Mechanics and Rock Engineering*, "Investigation of rock mass stability investigation around the tunnels in an underground mine in USA using three-dimensional numerical modeling," Y. Xing, P.H.S.W. Kulatilake & L. A. Sandbak, 2018a.

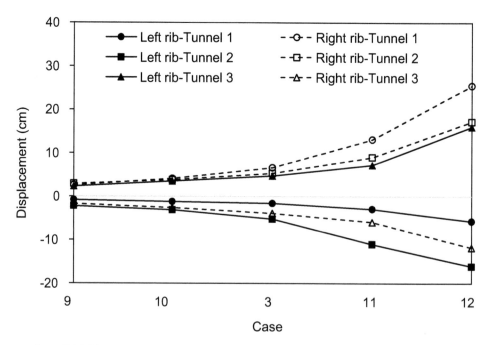

Figure 7.24 Maximum displacements on the ribs of the tunnels for Cases 3 and 9–12

Source: Modified from *Engineering Geology*, 238, Y. Xing, P.H.S.W. Kulatilake & L. A. Sandbak, "Effect of rock mass and discontinuity mechanical properties and delayed rock supporting on tunnel stability in an underground mine," pp. 62–75, 2018b, with permission from Elsevier.

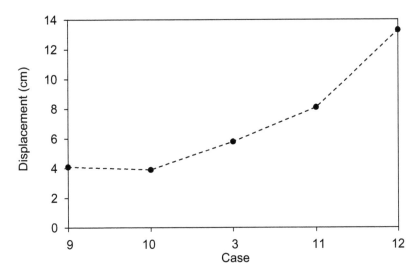

Figure 7.28 Maximum shear displacement on the major fault for Cases 3 and 9–12

Source: Modified from *Engineering Geology*, 238, Y. Xing, P.H.S.W. Kulatilake & L. A. Sandbak, "Effect of rock mass and discontinuity mechanical properties and delayed rock supporting on tunnel stability in an underground mine," pp. 62–75, 2018b, with permission from Elsevier.

Figures 7.30 and 7.31 show the deformations around the tunnels for the three cases. When the fault property values reduce, the maximum displacements of the rock masses increase slightly. A similar phenomenon can be observed for the failure zones around the tunnels as shown in Figures 7.32 and 7.33.

7.5.6   Evaluation of the applied support system

The procedures described in Section 7.3 were followed to apply the rock supports. Multiple stress relaxation factors, including the instantaneous supporting ($\lambda = 0$), are assigned to simulate the different delayed time of supporting, as given in Table 7.6. The effectiveness of the applied support system (Cases 17 and 29–32) is evaluated using the reductions of the deformations and of the failure zone thicknesses around tunnels from the unsupported situation (Case 3). Results are given in Figures 7.34 and 7.35. For the instantaneous supporting, the reduction of the maximum displacements is about 3%–8% on the ribs (Case 17 in Figure 7.34) and 2%–7% on the roofs and floors (Case 17 in Figure 7.35). As the installation is delayed at a stress relaxation of 20%, the deformations around the tunnels are well controlled, especially on the roofs and floors, with the maximum reduction up to 24% (Case 29 in Figure 7.35). As longer delayed periods are considered (Cases 30–32), the reductions of the deformation keep increasing with small fluctuations at several locations. For reductions of the failure area around the tunnels (Figures 7.36–7.37), more improvements can also be seen in the delayed cases (Cases 29–32) than in the instantaneous situation (Case 17). Yet the reductions are not distinct as that of the displacements; constant values can be observed at some locations. The results of maximum displacements and average failure zone thicknesses around the tunnels are provided in Tables B.1 and B.2 in Appendix B.

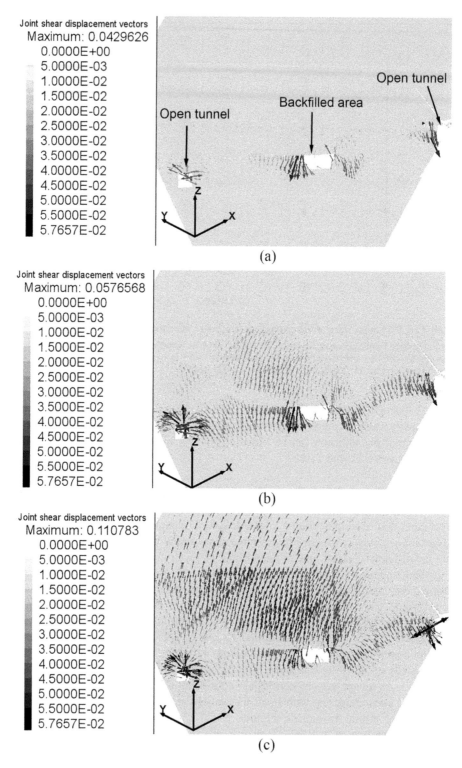

Figure 7.29 Joint shear displacement distributions on the major fault for (a) Case 13; (b) Case 3; (c) Case 14 (unit: m)

Source: Modified from *Engineering Geology*, 238, Y. Xing, P.H.S.W. Kulatilake & L. A. Sandbak, "Effect of rock mass and discontinuity mechanical properties and delayed rock supporting on tunnel stability in an underground mine," pp. 62–75, 2018b, with permission from Elsevier.

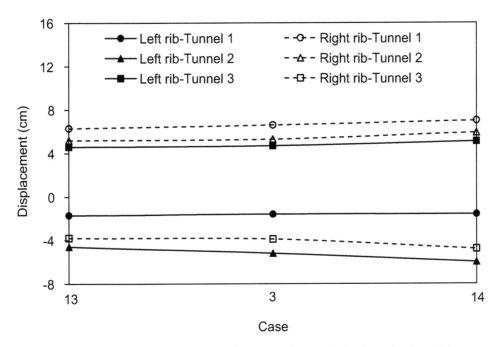

Figure 7.30 Maximum displacements on the ribs of the tunnels for Cases 3, 13 and 14

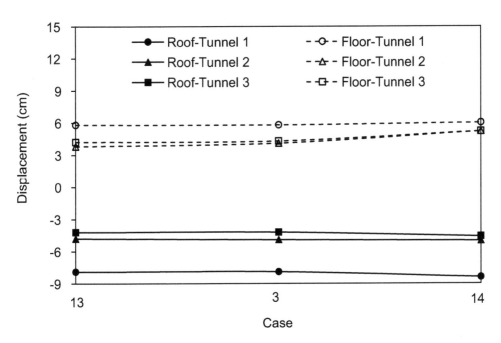

Figure 7.31 Maximum displacements on the roofs and the floors of the tunnels for Cases 3, 13 and 14

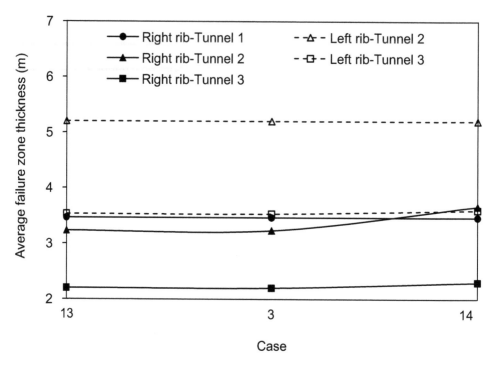

Figure 7.32 Average failure zone thicknesses on the ribs of the tunnels for Cases 3, 13 and 14

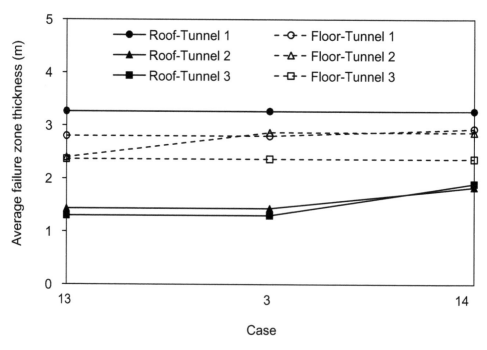

Figure 7.33 Average failure zone thicknesses on the roofs and the floors of the tunnels for Cases 3, 13 and 14

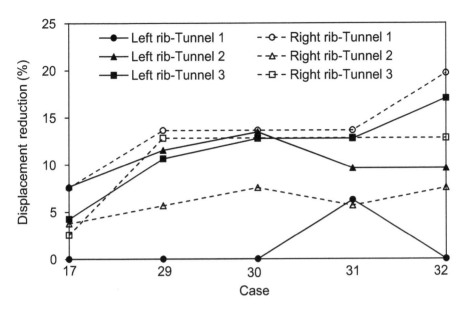

Figure 7.34 Reduction of the maximum displacements on the ribs of the tunnels for Cases 17 and 29–32

Source: Modified from *Engineering Geology*, 238, Y. Xing, P.H.S.W. Kulatilake & L. A. Sandbak, "Effect of rock mass and discontinuity mechanical properties and delayed rock supporting on tunnel stability in an underground mine," pp. 62–75, 2018b, with permission from Elsevier.

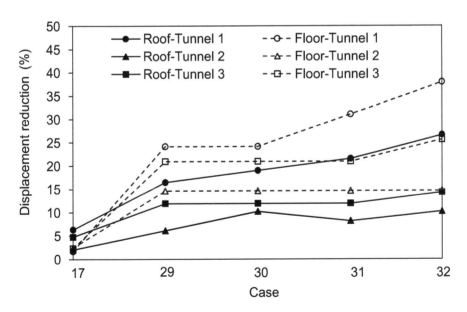

Figure 7.35 Reduction of the maximum displacements on the roofs and the floors of the tunnels for Cases 17 and 29–32

Source: Modified from *Engineering Geology*, 238, Y. Xing, P.H.S.W. Kulatilake & L. A. Sandbak, "Effect of rock mass and discontinuity mechanical properties and delayed rock supporting on tunnel stability in an underground mine," pp. 62–75, 2018b, with permission from Elsevier.

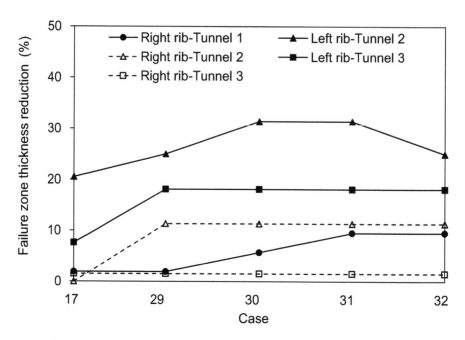

Figure 7.36 Reduction of the average failure zone thicknesses on the ribs of the tunnels for Cases 17 and 29–32

Source: Modified from *Engineering Geology*, 238, Y. Xing, P.H.S.W. Kulatilake & L. A. Sandbak, "Effect of rock mass and discontinuity mechanical properties and delayed rock supporting on tunnel stability in an underground mine," pp. 62–75, 2018b, with permission from Elsevier.

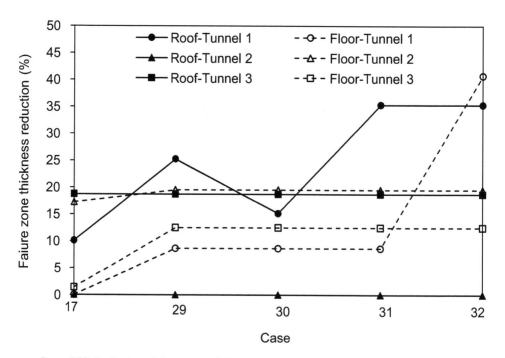

Figure 7.37 Reduction of the average failure zone thicknesses on the roofs and the floors of the tunnels for Cases 17 and 29–32

Source: Modified from *Engineering Geology*, 238, Y. Xing, P.H.S.W. Kulatilake & L. A. Sandbak, "Effect of rock mass and discontinuity mechanical properties and delayed rock supporting on tunnel stability in an underground mine," pp. 62–75, 2018b, with permission from Elsevier.

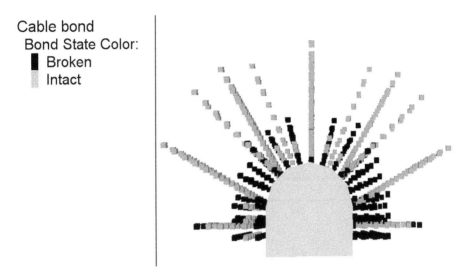

Figure 7.38 State of bond failure of the applied bolt supports in Case 17

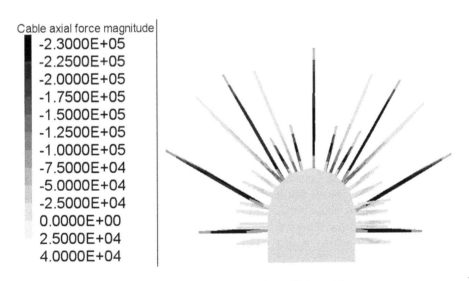

Figure 7.39 Axial force of the applied supports in Case 17 (unit: N)

The safety of the applied supports is analyzed through the bond shear failure and bolt tensile failure. Figure 7.38 shows the state of bond failure for a part of representative bolt supports for Case 17. It can be seen that the bond shear failure (in black cubes) is mostly taking place on the short bolts on the ribs (split sets) and the short ones on the roof (resin bolts). On the contrary, the bond of longer bolts either on the ribs (Swellex bolts) or on the roof (Swellex and cable bolts) are safe and shown as intact (in grey cubes). Figure 7.39 shows the axial force distribution of the supports. Unlike the bond failure, the longer bolts and the short roof bolts (resin bolts) have high axial forces, with the highest reaching

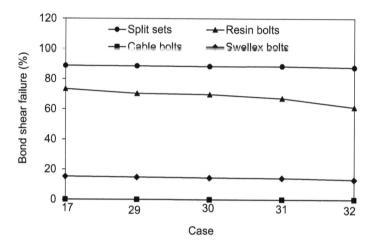

Figure 7.40 Bond shear failure of the supports for Cases 17 and 29–32

Source: Reprinted from *Engineering Geology*, 238, Y. Xing, P.H.S.W. Kulatilake & L. A. Sandbak, "Effect of rock mass and discontinuity mechanical properties and delayed rock supporting on tunnel stability in an underground mine," pp. 62–75, 2018b, with permission from Elsevier.

230 KN, while the forces of short wall bolts (split sets) are less than 60 KN. To evaluate the failure condition for the whole support system, the percentages of the bond shear failure and of the bolt tension failure were calculated based on the support information exported from the software. Figure 7.40 presents the results for Cases 17 and 29–32. Around 90% of the bond of split sets on the ribs can be seen as failing and 60%–75% of the bond of resin bolts on the roofs has failed, agreeing well with the situation illustrated in Figure 7.38. The bond failures are slightly reduced with the increasing delayed time, from Case 29 to Case 32. The cable and Swellex bolts are safe with respect to the bond condition. Figure 7.41 provides the percent of the bolts of which the axial forces have reached the tensile yield capacity. To be on the conservative side, the bolts having the axial force close to the tensile yield capacity within 1 KN are considered as failing. It seems that none of the split sets and a small amount of the cable bolts have failed in tension; about 20%–35% of the resin bolts and 50%–60% of the Swellex bolts have failed, fitting what is shown in Figure 7.39. Again, the number of failing bolts reduces slightly with the increasing delayed time.

It can be concluded that by applying the rock supports, the deformations and failure zones around the tunnels have been controlled to a certain degree, and the reductions were further improved by the delayed supporting. In addition, the improvements seemed to be more on the roofs and floors than on the ribs, where denser and stronger rock supports were applied. The extended delayed installation also slightly improves the safety of the supports. It indicates that the installation time of rock supports with respect to the excavation is critical to the support performance. Allowing a certain amount of rock deformations after tunnel excavation could enhance the effectiveness of rock supports. On the other hand, too late installation of the supports probably also leads to ineffective supporting, which may be connected to the fluctuations presented in Figures 7.34 through 7.37. With respect to the safety condition of the applied rock supports, a large amount of the bond of split sets and resin bolts has failed; some resin and Swellex bolts underwent tensile failure. It may imply that the bond

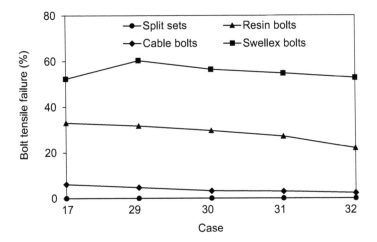

Figure 7.41 Bolt tensile failure of the supports for Cases 17 and 29–32

Source: Reprinted from *Engineering Geology*, 238, Y. Xing, P.H.S.W. Kulatilake & L. A. Sandbak, "Effect of rock mass and discontinuity mechanical properties and delayed rock supporting on tunnel stability in an underground mine," pp. 62–75, 2018b, with permission from Elsevier.

and bolt tension strengths of the supports are insufficient. Considering the average failure zone thickness of 2–4 m at ribs and of 1.3–3.0 m on the roofs for the supported cases given in Table B.2, and the fact that rapid movements are located within a distance of 2 or 3 m from the tunnel surface, the split sets on the walls (length = 1.83 m) and the resin bolts on the roofs (length = 2.44 m) might be short in length.

It should be mentioned that the above-presented support failures (Figures 7.38–7.41) are for the three installations of rock supports, as described in Section 2.5.1 and Figure 2.4. They are the initial supports applied for the rock masses. In the field, rehabilitation works had been conducted by installing longer and stronger supports from time to time, as described in Section 2.5.4. For example, the failed split sets have been replaced using the 2.44 m Swellex or inflatable bolts, which are quick to install and are more resistance to shear; additional repair rehabilitation was applied using the 2.44 m and 3.66 m Swellex bolts. These rehabilitation works were not included in the numerical model. In practice, the shotcrete and wire mesh were applied (see Section 2.5.2), which also helped to improve the rock mass and bolt conditions. Therefore, the predicted failure conditions of the rock supports are higher than that in the field; discrepancies are likely to exist in the failure modes of the supports in the numerical model and that in the field. The numerical results shown in Figures 7.40 and 7.41 seem to be the worst case scenario rather than the actual percentage of support break in the field. The ground control problems at the mine are not instantaneous, but a slow deterioration over time and a low creep of weak rock for low RMR rocks. Major field rehabilitation by the addition of ground support happens every five years or so, based on cracking and movement history. A few bolts can fail, usually in weak, wet or faulted zones, and may initiate unraveling ground past the bolt. It also demonstrates that the failure (break or plate failure) is occurring, and rehabilitation is necessary to re-establish stability, and to prevent ground falls.

Table 7.7 Comparisons between the field measurements and numerical modeling results

Case No.	Convergence strain at location MT-17 (%)	Convergence strain at location MT-18 (%)
Field measurement	1.96	1.27
15	2.23	1.29
16	2.25	1.03
17	2.01	1.20
18	2.21	1.29
19	2.79	1.58
20	4.26	2.09
21	3.15	2.19
22	1.16	0.67
23	0.96	0.55
24	1.31	0.79
25	3.41	2.25
26	6.55	3.71
27	2.36	1.33
28	1.94	1.12
29	1.96	1.13
30	1.94	1.09
31	1.97	1.09
32	1.92	1.06
33	1.94	1.08

7.5.7 Comparisons between the numerical modeling results and field measurements

The numerical modeling results are compared with the field measurements from tape extensometers at two locations, MT-17 and MT-18 shown in Figure 2.8. Due to the geometry difference between the tunnels built in the numerical model and that in the field, the convergence strain (tunnel convergence over tunnel dimension) is utilized. Table 7.7 summarizes the results for Cases 15–33 and of the field measurements. It shows that the obtained strains for Cases 15–18, and 27–33 are in good agreement with the field measurements. They are the situations with K_0 values of 0.5–1.25, with average rock mass condition, and with the support system (see Table 7.6). The results of the cases with different fault property values (Cases 17, 27 and 28) show insignificant differences compared to the field measurements; the ranges are respectively 1.94%–2.36% at MT-17 (note the field value is 1.96%) and 1.12%–1.33% at MT-18 (note the field value is 1.27%). The cases with high K_0 of 1.5 and 2.0 (Cases 19 and 20), with low RMR values (Case 21) and with low residual strengths (Cases 25–26), have much greater strain values than that of the field measurements. Conversely, the strain values of the cases with high RMR values (Case 22) and high residual strengths (Cases 23–24) are less than that of the field measurements.

7.5.8 Discussions on the design of rock supports

As described in Section 7.5.1, the presence of the dike interlayer, the fault, and many mine construction activities have posed enormous influence on the stability of nearby tunnels. Therefore, to assist the design of appropriate rock supports, these factors should be taken

Table 7.8 Maximum extent of strength degradation area around the tunnels on vertical planes of X = −30 m, X = 0 m, X = 27 m, Y = 20 m and Y = −20 m[a]

Tunnel cross section	Roof (m)	Floor (m)	Left wall (m)	Right wall (m)
X = −30 m	2.81 extended to Dike	2.99	N/A extended to the backfilled area	4.02
X = 0 m (L)	4.38	3.38	N/A backfilled	4.27
X = 0 m (R)	3.48	4.24	7.46	4.92
X = 27 m	4.37	4.60	4.22	4.52
Y = 20 m (L)	5.54 extended to Dike	3.75	N/A backfilled	N/A connecting right-side tunnel
Y = 20 m (R)	3.34	3.38	N/A connecting left-side tunnel	4.41
Y = −20 m	3.71	3.49	4.36	4.33

[a] Data from Xing et al. (2019).

into account. Based on the comparisons given in Table 7.7, supplementary results are provided and analyzed for the case with the K_0 value of 1.0, with the average rock mass and fault property values, and with the stress relaxation ratio of 50% (Case 33 in Table 7.6). Table 7.8 provides the maximum extents of shear strength (cohesion) degradation area around the tunnels on the vertical planes of X = −30 m, X = 0 m, X = 27 m, Y = 20 m and Y = −20 m (see Figure 7.5). To understand the results well, it is better to refer to the diagrams in Figure 7.12. For the roofs of various tunnels, the maximum extents of degradation area are in the range of 2.8 m and 5.6 m (Table 7.8). On the plane of X = −30 m, the extent of 2.81 m represents the vertical distance above the crown, which has extended to the dike layer. Whereas Figure 7.12a showed that the degradation area extends to the dike deeply on the upper-right side. Under such a condition, the bolt length is suggested to be long enough to penetrate the dike layer (probably 5.3 m) to make the rock mass stable. The peak value of 5.54 m appeared on the roof of the left-side tunnel on the plane of Y = 20 m (see Table 7.8). Yet the immediate roof has been excavated and backfilled (see Figure 7.12d), of which the movement seemed to be controlled so far, as mentioned in Section 7.5.1 and illustrated in Figure 7.11d (Roof-L). Attention probably needs to be paid to this location after a certain period of time, in months or years, in case of the failing of backfilling materials. The roofs of the left tunnel on the plane of X = 0 m and of the tunnel intersection on the plane of X = 27 m have the strength degradation area extents of about 4.4 m. For the rest tunnels, the roofs are in relatively better conditions, with an approximate value of 3.5 m (see Table 7.8). Typical thicknesses of the strength degradation area on the floors are 3–4.6 m. For the ribs, most of the strength degradations are between 4 and 5 m. Distinguishable value of 7.46 m is observed on the left wall of the right-side tunnel on the plane of X = 0 m, which can be explained by the close distance to the major fault (see Figure 7.12b). It can be seen from Table 7.8 that results are not available at several locations because of the extension to the backfilled area or adjacent tunnel. For instance, the left area to the tunnels on the planes of X = −30 m and X = 0 m was excavated and backfilled, and the failure covers a long distance. It does not mean that these places are unstable; in fact, the failures may take place before the backfilling, and the rock mass stability has been improved afterward. Nevertheless, the two

Table 7.9 Guidelines for support of rock tunnels in accordance with the rock mass rating system[a,b]

Rock mass class	Rock bolts (20 mm diameter, fully grouted)	Shotcrete	Steel sets
Very good rock RMR: 81–100	Generally, no support required except for occasional spot bolting		
Good rock RMR: 61–80	Locally, bolts in crown 3 m long, spaced 2.5 m, with occasional wire mesh	50 mm in crown where required	None
Fair rock RMR: 41–60	Systematic bolts 4 m long, spaced 1.5–2 m in crown and walls with wire mesh in crown	50–100 mm in crown and 30 mm in sides	None
Poor rock RMR: 21–40	Systematic bolts 4–5 m long, spaced 1–1.5 m in crown and wall with wire mesh	100–150 mm in crown and 100 mm in sides	Light to medium ribs spaced 1.5 m where required
Very poor rock RMR: <20	Systematic bolts 5–6 m long, spaced 1–1.5 m in crown and walls with wire mesh. Bolt invert	150–200 mm in crown, 150 mm in sides, and 50 mm on face	Medium to heavy ribs spaced 0.75 m with steel lagging and forepoling if required. Close invert

[a] Data from Z. T. Bieniawski, *Engineering Rock Mass Classifications: a complete manual for engineers and geologists in mining, civil, and petroleum engineering* (John Wiley & Sons, 1989), pp. 62.

[b] Shape: horseshoe; width: 10 m; vertical stress: < 25 MPa; construction: drilling and blasting

connecting ribs on the plane of Y = 20 m seem to suffer unfavorable situations due to extensive failures taking place in-between, including the dike layer. For such a location, special ground control measures (extremely long cable bolts, shotcrete arch, etc.) may be needed.

In conclusion, based on the supplementary results of Case 33 and those presented in previous sections, the bolt lengths on the roof and on the walls are suggested to be 3–4 m and 4–5 m, respectively. Longer and denser bolts, or even other stronger reinforcements, should be applied at some locations when unfavorable situations are encountered, such as the occurrence of the weak rock material or a fault and the frequent and adjacent engineering construction activities. Specific designs can be made combining numerical results with field observations.

Bieniawski (1989) provides guidelines for selecting supports based on the RMR values, as given in Table 7.9. The rock masses having the RMR values of 32.5 and 40 (the average values in the present study) are classified as the "poor rock." For the excavations in such rock masses (21 < RMR < 40), the empirical support guidelines suggest fully grouted rock bolts of length 4–5 m at a spacing of 1–1.5 m in the crown and walls. Although the suggestions are applied for the tunnels with a width of 10 m, it indicates that the above conclusions based on the numerical results are reliable.

Chapter 8

Summary, conclusions and recommendations for future work

8.1   Summary and conclusions

This monograph presents a comprehensive case study on the investigation of rock mass stability around tunnels for an underground mine in the USA. The following major conclusions are made.

The geological conditions at the selected mine site are complex, including the inclined ore bodies, many fault zones and an intrusive dike. Geological surveys showed that the rock mass conditions are in wide ranges where the rock masses in ore zones have low RMR values. The targeted tunnel system is spatially complicated, which may go through the low rock mass quality areas or major faults. Field surveys to the site showed that the rock masses at some locations (i.e. tunnel intersections, weak rock areas and fault zones, suffered high deformations and failures). In the field, rehabilitation work has been conducted repeatedly from time to time to keep the rock mass stable. For the tunnels located in the very weak ground, the shotcrete arch support system was designed. Since no in-situ stress measurement was conducted, an empirical equation was used by the mining company to estimate the lateral stress ratio at the mine, and the values have turned out to be between 0.5 and 1.0.

Compared to the rock mass classifications, analytical methods and field instrumentation methods, the numerical method is a low-cost but highly efficient approach, which is capable of revealing rock mass failure mechanisms for rock engineering problems that have complex geometries, geological conditions and multiple impact factors. The distinct feature of discontinuum modeling is to represent the discontinuities explicitly so that large displacements and rotations, including total detachment, of the discrete blocks are allowed. In Chapter 3, some DEM codes and their applications were reviewed; also discussed are the strengths and limitations of each code. For a specific problem, it is important to be aware of its particular characteristics and the objectives, and hence to develop an adequate model to capture the most relevant factors.

Some critical aspects pertinent to the modeling of underground rock problems include the representation of joint features (i.e. the joint geometry patterns and joint material properties, the estimation of equivalent rock mass properties and selection of rock mass constitutive models and the simulation of in-situ stress and construction sequence).

Laboratory tests were carried out on the LM and ML rocks containing limestone and mudstone, and the dacite. Close physical and mechanical property values were obtained for the two former rocks, which are higher than those of the dacite rock. For smooth joints, the nonlinear variation of joint stiffnesses was obtained in relation to normal stress. Small to moderate variations were observed in the test results. Due to the fact that the rock samples were taken from widely spatially distributed boreholes, the grain components and the number and distribution of fractures could be very different from sample to sample. Besides, the unavoidable man-made errors or unexpected failure of the sample (hydrostone cast in the direct shear test, for example) also led to uncertainties of the results.

Based on the available geological, geotechnical and mine construction information, a sophisticated three-dimensional discontinuum model was built using the 3DEC. The model incorporates the complicated geologies including a weak inclined dike layer and a few large-scale faults, a complicated tunnel system, and the conducted backfilling activities at the mine. The effect of minor discontinuities is implicitly included by combining with intact rock properties. Sequential construction and delayed supporting procedures were implemented numerically. Stress analyses were performed to evaluate the impacts of various factors on tunnel stability.

Numerical results showed that the complicated geologies (i.e. the faults and weak dike layer) and the engineering construction activities have posed large influences on the rock mass behavior around tunnels, such as the asymmetric distributions and high values of the stresses and displacements, as well as more failure zones. The movements of interior rock masses indicated that relatively large displacements occurred within a distance of 2 or 3 m from the tunnel surface and the values reduced to certain constant values after 4–5 m away.

The lateral stress ratio was varied to investigate the effect of horizontal in-situ stress on tunnel stability. As K_0 increases, the maximum major principal stress rotates from the ribs to the roofs and floors; the value decreased at first but then increased. Both horizontal and vertical convergences of the tunnels increased with the increasing K_0 with small fluctuations. The failure zone thicknesses on the ribs increased slightly from $K_0 = 0.5$ to $K_0 = 2.0$, while increased significantly on the roofs and floors. The shear behavior of the major fault was greatly affected by K_0 values, in both the magnitude and direction. In general, the tunnels in the cases with $K_0 = 1.0$ are the most stable, with the lowest stresses and small deformations along the fault; $K_0 = 1.5$ and $K_0 = 2.0$, however, are the most unstable cases having severe roof and floor problems.

By varying the RMR values, three cases with different rock mass property values were analyzed. Intuitive results have been obtained; the soft system is the most unstable case, while the stiff system caused the least deformations and failure zones. Additionally, the soft case resulted in more deformations and failures for the two tunnels on the plane with unfavorable geologies and more disturbances.

The impact of the residual strength can be seen from the great changes of the deformations and failure zones around the tunnels. The instability of tunnels was much sensitive to low residual strengths. The growth of the movement along the fault was observed near the open and backfilled excavations.

The varying fault properties largely changed the magnitudes of the shear displacement along the major fault. High values can be observed near the backfilled or open excavations. Nevertheless, the rock mass behavior was also slightly affected.

The applied support system improved the stability of the tunnels, and the deformations and failures were further reduced by considering the delayed supporting. Failure conditions of the rock supports showed that a large amount of the bonds of short bolts on the ribs and on the roof failed, while the longer Swellex and cable bolts tended to fail in bolt tension. When the delayed time was increased, the failure condition of the bolts improved slightly; the reductions of the deformations and failures of the rock masses varied with small fluctuations. It might indicate that the delayed time is critical to the function of the supports and interactions between the supports and the rock masses. With respect to the support failure condition, it may represent the worst situation in the field for lack of incorporating the shotcrete and mesh supports, the frequent rehabilitation works, etc. in the numerical model. According to the rock mass behavior, the bolt length was suggested to be 3–4 m on the roof and 4–5 m on the walls. Longer and denser bolts, or ever other stronger reinforcements should be applied at some locations when unfavorable situations are encountered, such as the weak rock material or a fault and the frequent and adjacent engineering construction activities. Specific designs can be made combining numerical results with field observations. The suggestions of bolt length were in accord with the guidelines provided by Bieniawski (1989) for selecting the supports based on the rock mass ratings.

Finally, numerical modeling results were compared with the field measurements at two locations. The cases where the rock masses are in average rock mass condition and in average residual strength and with K_0 values in the range of 0.5–1.25 provide well-matched results. The predictions of the other cases are in reasonable ranges.

The supplemental laboratory test results and numerical model results that helped to illustrate the analysis in previous chapters are given in Appendices A and B.

8.2   Recommendations for future work

As summarized above, an elaborate study was carried out to investigate the tunnel stability for the underground mine. Nevertheless, if more resources and time are given, improvements or further work can be performed on the following aspects:

1   To provide more reliable input parameters for the rock masses and faults, it is suggested to collect rock samples and the samples containing natural joints from or close to the selected study area. The rock mass residual strength can be determined using the indirect methods based on GSI values, which can be obtained from field observations of the rock mass condition.

2   The minor discontinuities can be explicitly represented in the numerical model. However, this will lead to a large amount of computational time for a three-dimensional model including complex features as presented in this research, and probably cause numerical instabilities. Careful considerations or trial calculations may be needed in the first place.

3   Since the failure zone thickness is sensitive to the element size, finer meshes for the near field can be used to provide more precise results. Yet, this will also result in an increase of computational time.

4   Sensitivity analyses could be performed for the rock supports by varying bolt length, bond strength, bolt diameter, bolt tensile yield capacity and bolt spacing in certain ranges. Simulation of other supports applied at the mine, such as shotcrete and steel sets, may be needed. The installation time of the supports can be determined by performing back analysis from field deformation measurements.

Appendix A

Supplementary of laboratory test results

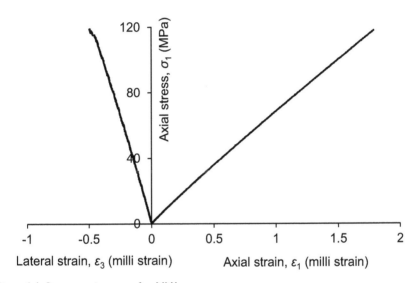

Figure A.1 Stress-strain curve for MU1

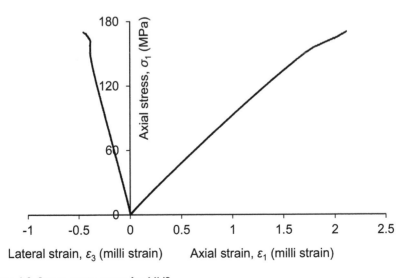

Figure A.2 Stress-strain curve for MU3

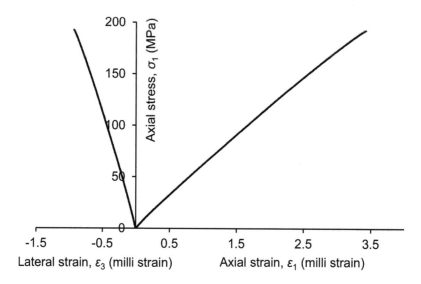

Figure A.3 Stress-strain curve for LU I

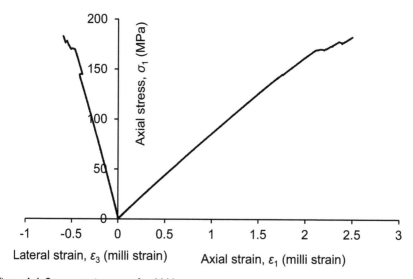

Figure A.4 Stress-strain curve for LU4

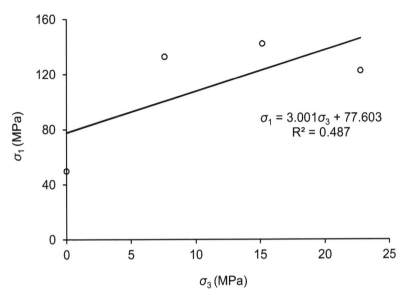

$\sigma_1 = 3.001\sigma_3 + 77.603$
$R^2 = 0.487$

Figure A.5 Linear regression of triaxial test results for dacite

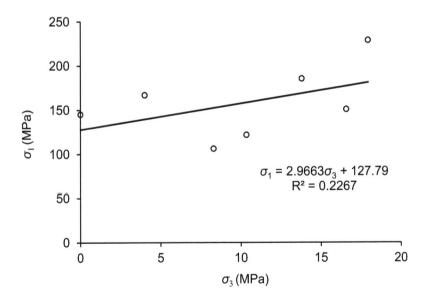

$\sigma_1 = 2.9663\sigma_3 + 127.79$
$R^2 = 0.2267$

Figure A.6 Linear regression of triaxial test results for ML

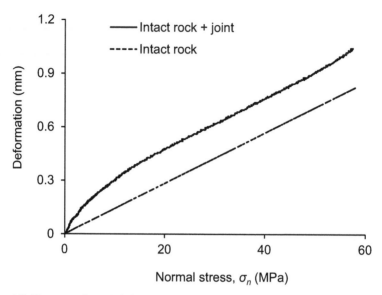

Figure A.7 Diagram of total deformation versus normal stress and intact rock deformation versus normal stress for DJKN2

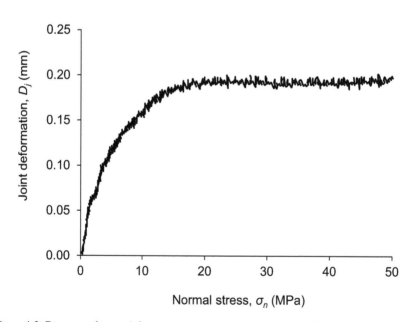

Figure A.8 Diagram of joint deformation versus normal stress for DJKN2

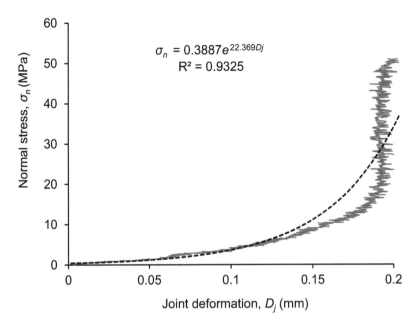

$$\sigma_n = 0.3887e^{22.369D_j}$$
$$R^2 = 0.9325$$

Figure A.9 Diagram of normal stress versus joint deformation and the fitted exponential curve for dacite joint (DJKN2)

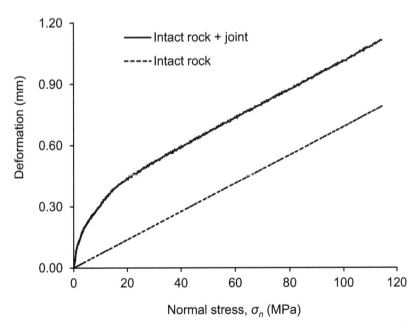

Figure A.10 Diagram of total deformation versus normal stress and intact rock deformation versus normal stress for MJKN1

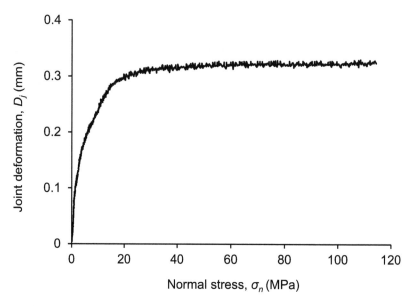

Figure A.11 Diagram of joint deformation versus normal stress for MJKN1

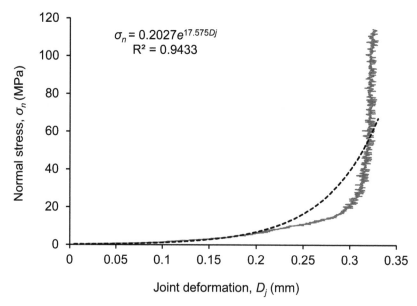

$$\sigma_n = 0.2027e^{17.575Dj}$$
$$R^2 = 0.9433$$

Figure A.12 Diagram of normal stress versus joint deformation and the fitted exponential curve for mudstone joint (MJKN1)

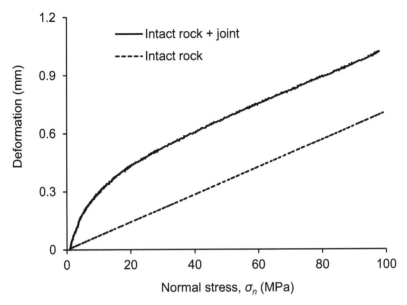

Figure A.13 Diagram of total deformation versus normal stress and intact rock deformation versus normal stress for MJKN2

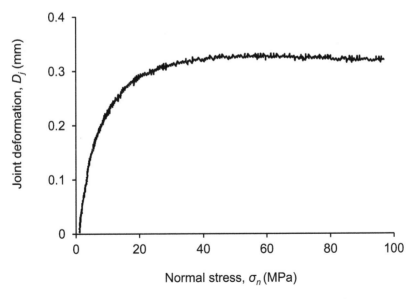

Figure A.14 Diagram of joint deformation versus normal stress for MJKN2

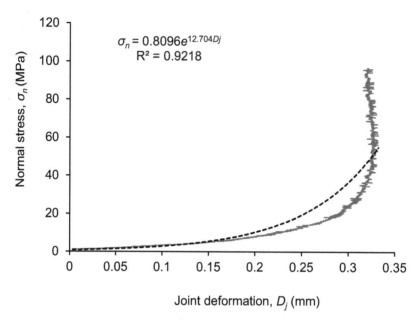

$$\sigma_n = 0.8096e^{12.704Dj}$$
$$R^2 = 0.9218$$

Figure A.15 Diagram of normal stress versus joint deformation and the fitted exponential curve for mudstone joint (MJKN2)

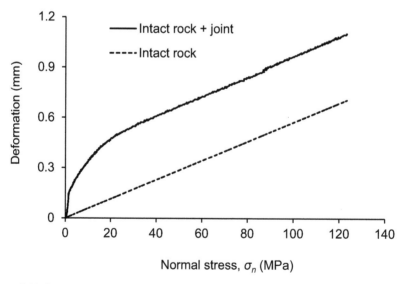

Figure A.16 Diagram of total deformation versus normal stress and intact rock deformation versus normal stress for MJKN3

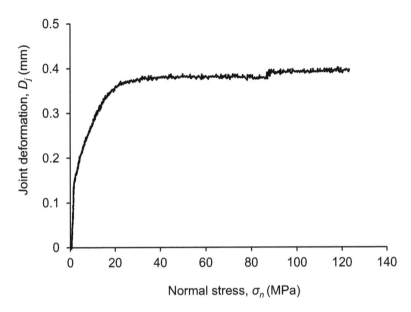

Figure A.17 Diagram of joint deformation versus normal stress for MJKN3

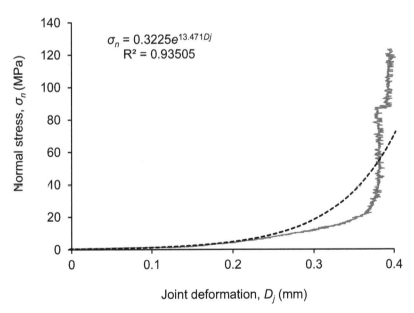

$$\sigma_n = 0.3225e^{13.471Dj}$$
$$R^2 = 0.93505$$

Figure A.18 Diagram of normal stress versus joint deformation and the fitted exponential curve for mudstone joint (MJKN3)

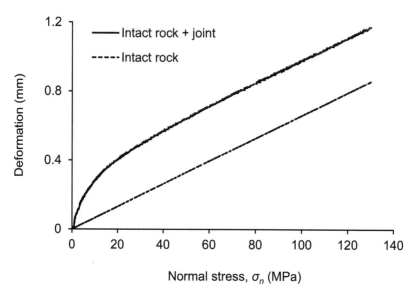

Figure A.19 Diagram of total deformation versus normal stress and intact rock deformation versus normal stress for LJKN1

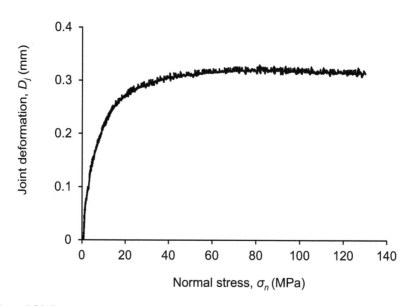

Figure A.20 Diagram of joint deformation versus normal stress for LJKN1

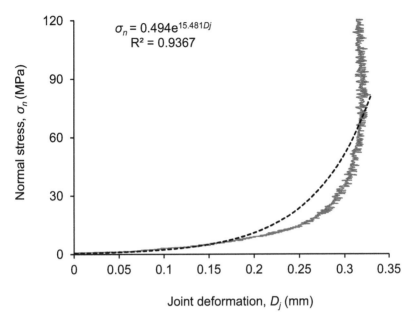

$$\sigma_n = 0.494e^{15.481Dj}$$
$$R^2 = 0.9367$$

Figure A.21 Diagram of normal stress versus joint deformation and the fitted exponential curve for mudstone joint (LJKN1)

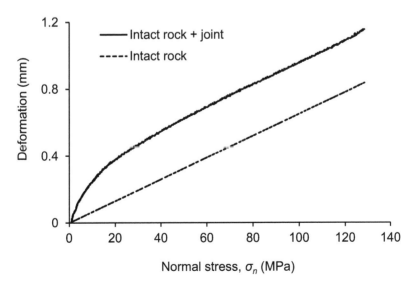

Figure A.22 Diagram of total deformation versus normal stress and intact rock deformation versus normal stress for LJKN2

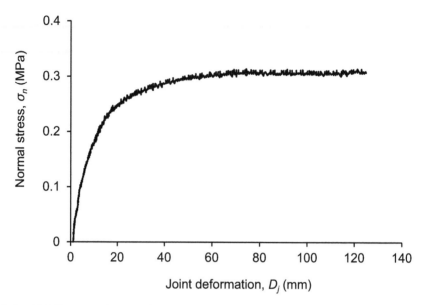

Figure A.23 Diagram of joint deformation versus normal stress for LJKN2

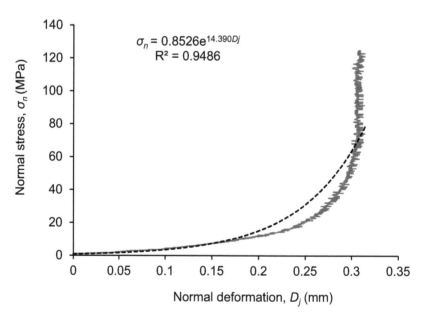

$$\sigma_n = 0.8526e^{14.390Dj}$$
$$R^2 = 0.9486$$

Figure A.24 Diagram of normal stress versus joint deformation and the fitted exponential curve for mudstone joint (LJKN2)

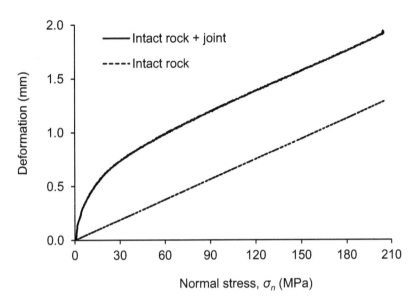

Figure A.25 Diagram of total deformation versus normal stress and intact rock deformation versus normal stress for LJKN3

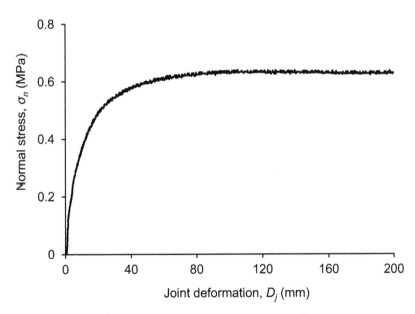

Figure A.26 Diagram of joint deformation versus normal stress for LJKN3

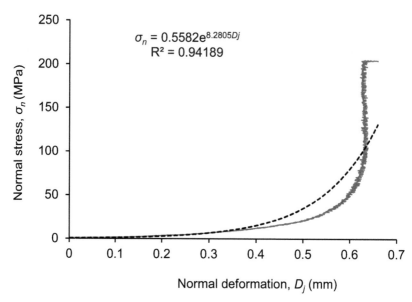

Figure A.27 Diagram of normal stress versus joint deformation and the fitted exponential curve for mudstone joint (LJKN3)

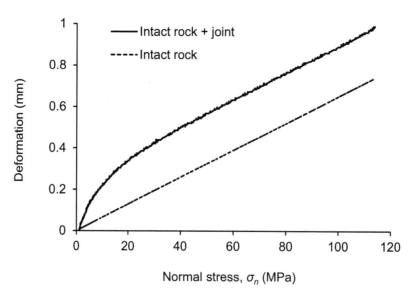

Figure A.28 Diagram of total deformation versus normal stress and intact rock deformation versus normal stress for DLJKN

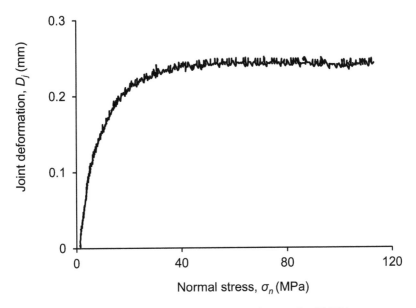

Figure A.29 Diagram of joint deformation versus normal stress for DLJKN

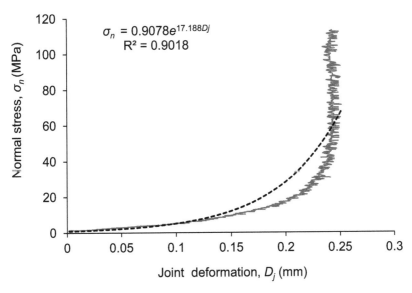

$$\sigma_n = 0.9078e^{17.188Dj}$$
$$R^2 = 0.9018$$

Figure A.30 Diagram of normal stress versus joint deformation and the fitted exponential curve for dacite-limestone interface (DLJKN)

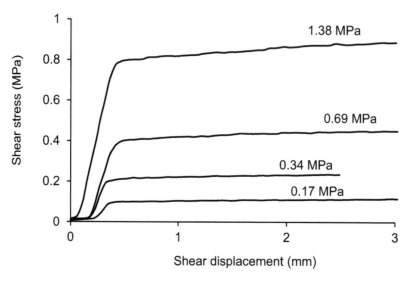

Figure A.31 Diagram of shear stress versus shear displacement curve for dacite joint (DS-2)

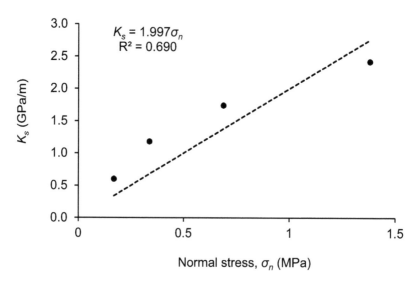

Figure A.32 Diagram of K_s versus normal stress and the fitted regression curve for dacite joint (DS-2)

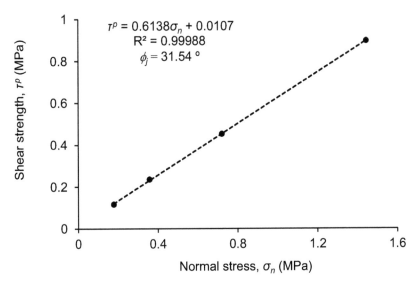

Figure A.33 Diagram of shear strength versus normal stress and the fitted regression curve for dacite joint (DS-2)

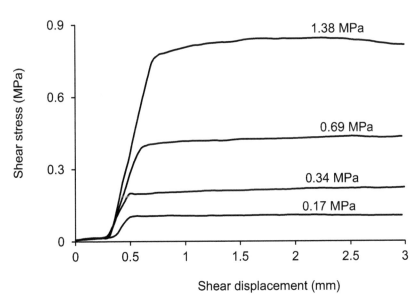

Figure A.34 Diagram of shear stress versus shear displacement curve for mudstone joint (MS-1)

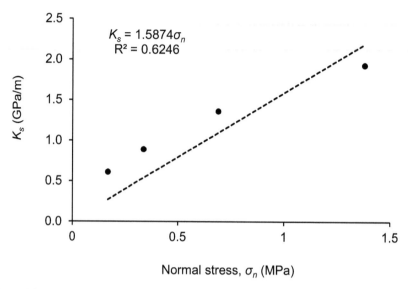

Figure A.35 Diagram of K_s versus normal stress and the fitted regression curve for mudstone joint (MS-1)

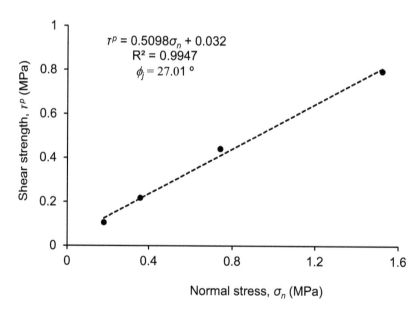

Figure A.36 Diagram of shear strength versus normal stress and the fitted regression curve for mudstone joint (MS-1)

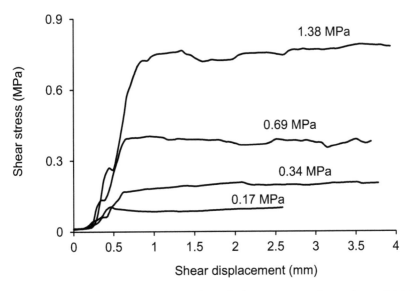

Figure A.37 Diagram of shear stress versus shear displacement curve for mudstone joint (MS-2)

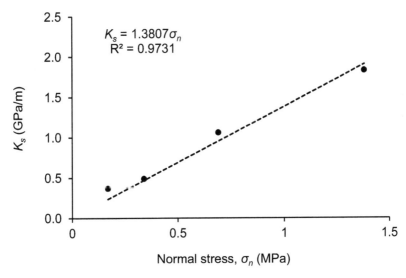

Figure A.38 Diagram of K_s versus normal stress and the fitted regression curve for mudstone joint (MS-2)

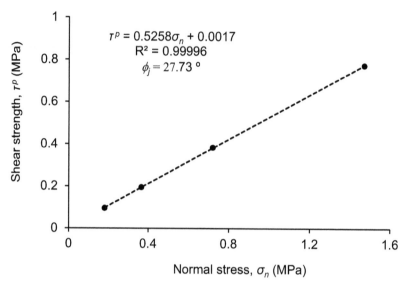

Figure A.39 Diagram of shear strength versus normal stress and the fitted regression curve for mudstone joint (MS-2)

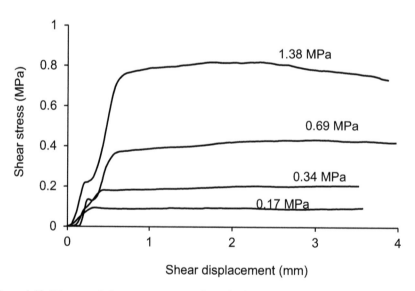

Figure A.40 Diagram of shear stress versus shear displacement curve for mudstone joint (MS-3)

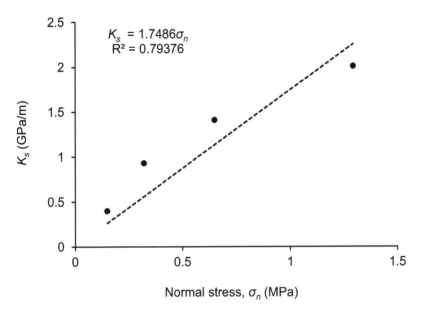

Figure A.41 Diagram of K_s versus normal stress and the fitted regression curve for mudstone joint (MS-3)

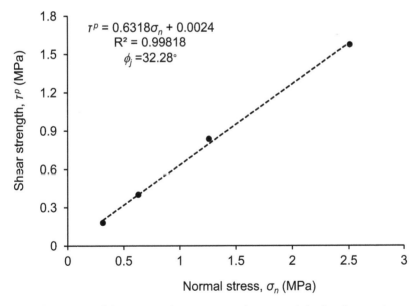

Figure A.42 Diagram of shear strength versus normal stress and the fitted regression curve for mudstone joint (MS-3)

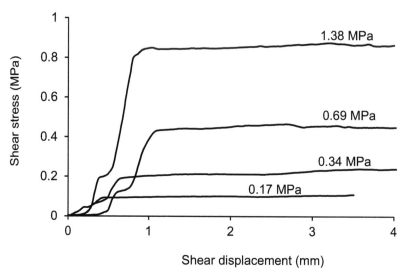

Figure A.43 Diagram of shear stress versus shear displacement curve for mudstone joint (MS-4)

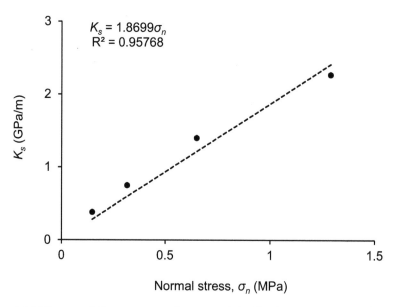

Figure A.44 Diagram of K_s versus normal stress and the fitted regression curve for mudstone joint (MS-4)

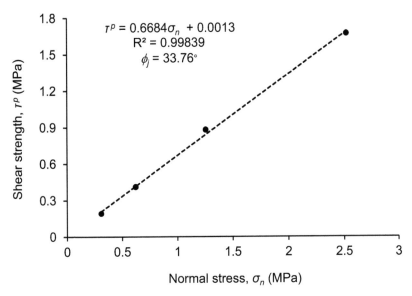

Figure A.45 Diagram of shear strength versus normal stress and the fitted regression curve for mudstone joint (MS-4)

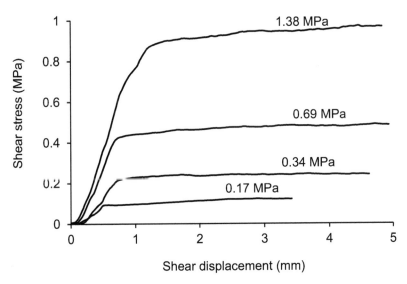

Figure A.46 Diagram of shear stress versus shear displacement curve for mudstone joint (MS-5)

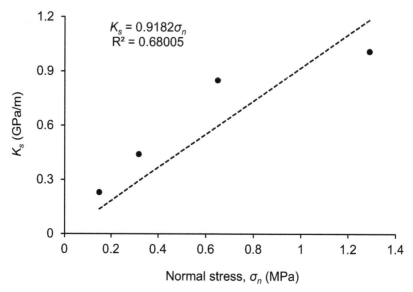

Figure A.47 Diagram of K_s versus normal stress and the fitted regression curve for mudstone joint (MS-5)

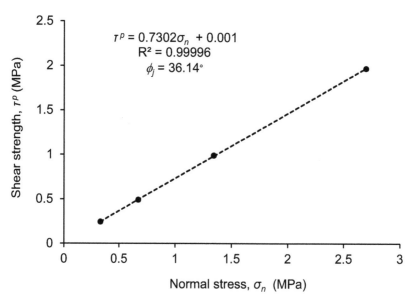

Figure A.48 Diagram of shear strength versus normal stress and the fitted regression curve for mudstone joint (MS-5)

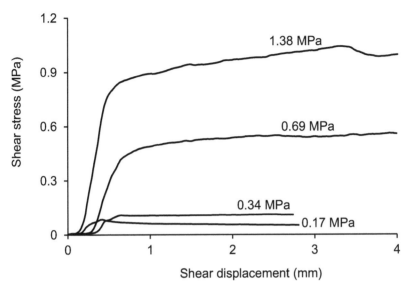

Figure A.49 Diagram of shear stress versus shear displacement curve for limestone joint (LS-1)

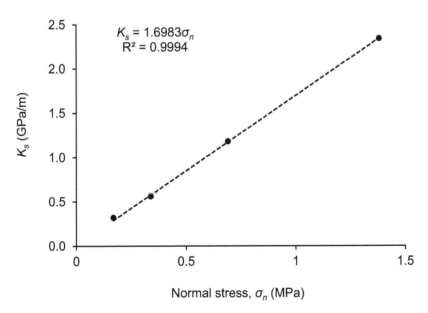

Figure A.50 Diagram of K_s versus normal stress and the fitted regression curve for limestone joint (LS-1)

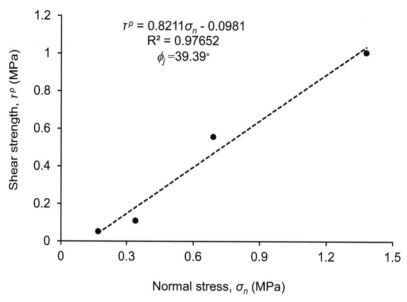

Figure A.51 Diagram of shear strength versus normal stress and the fitted regression curve for limestone joint (LS-1)

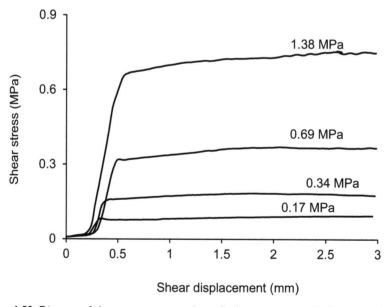

Figure A.52 Diagram of shear stress versus shear displacement curve for limestone joint (LS-2)

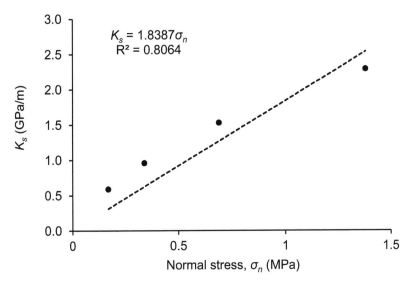

Figure A.53 Diagram of K_s versus normal stress and the fitted regression curve for limestone joint (LS-2)

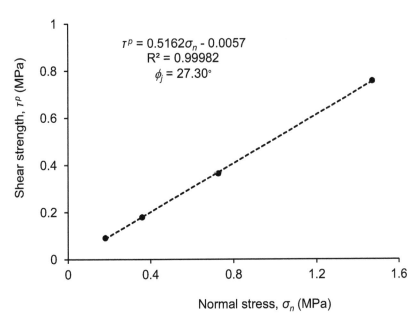

Figure A.54 Diagram of shear strength versus normal stress and the fitted regression curve for limestone joint (LS-2)

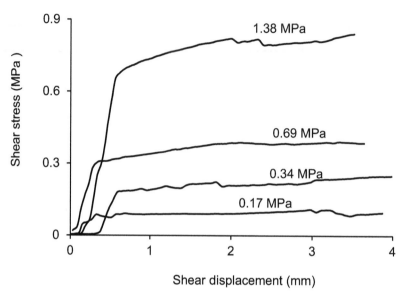

Figure A.55 Diagram of shear stress versus shear displacement curve for limestone joint (LS-3)

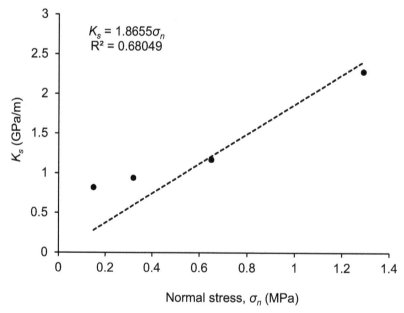

Figure A.56 Diagram of K_s versus normal stress and the fitted regression curve for limestone joint (LS-3)

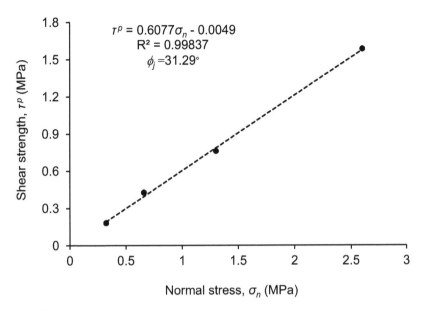

Figure A.57 Diagram of shear strength versus normal stress and the fitted regression curve for limestone joint (LS-3)

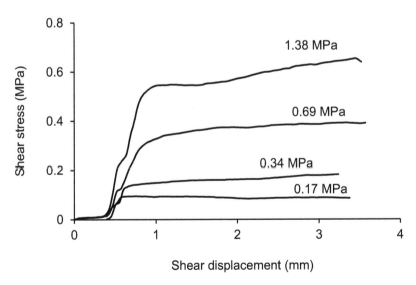

Figure A.58 Diagram of shear stress versus shear displacement curve for limestone joint (LS-4)

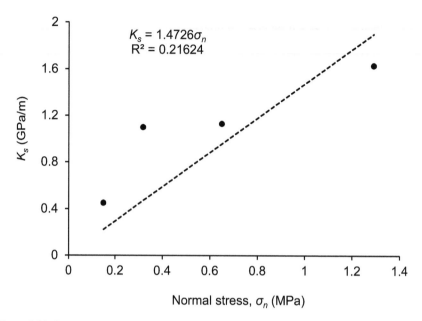

Figure A.59 Diagram of K_s versus normal stress and the fitted regression curve for limestone joint (LS-4)

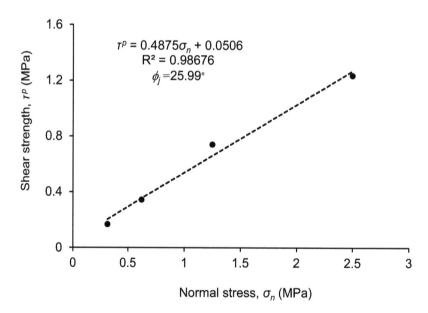

Figure A.60 Diagram of shear strength versus normal stress and the fitted regression curve for limestone joint (LS-4)

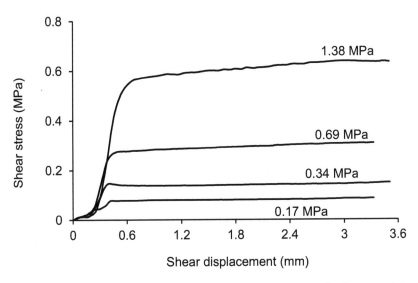

Figure A.61 Diagram of shear stress versus shear displacement curve for limestone joint (LS-5)

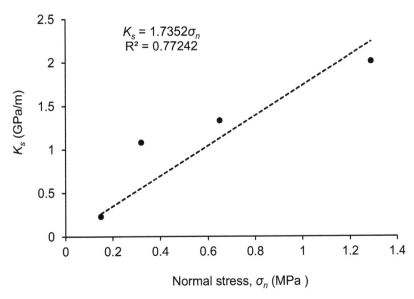

Figure A.62 Diagram of K_s versus normal stress and the fitted regression curve for limestone joint (LS-5)

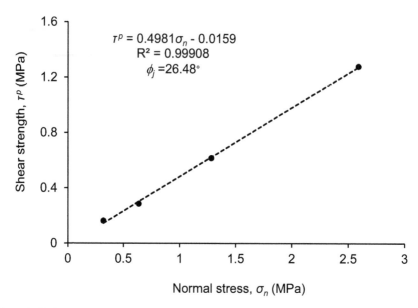

Figure A.63 Diagram of shear strength versus normal stress and the fitted regression curve for limestone joint (LS-5)

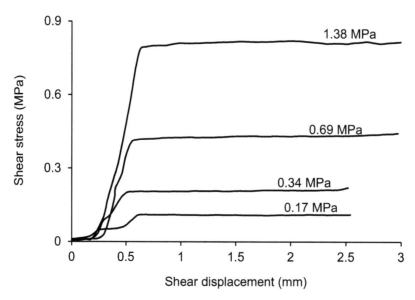

Figure A.64 Diagram of shear stress versus shear displacement curve for the interface between dacite and limestone (LDS-1)

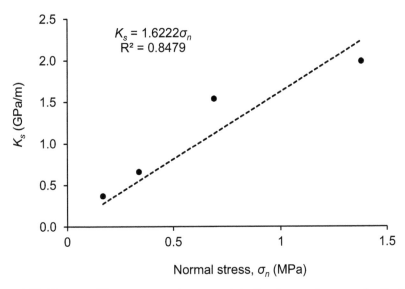

Figure A.65 Diagram of K_s versus normal stress and the fitted regression curve for the interface between dacite and limestone (LDS-1)

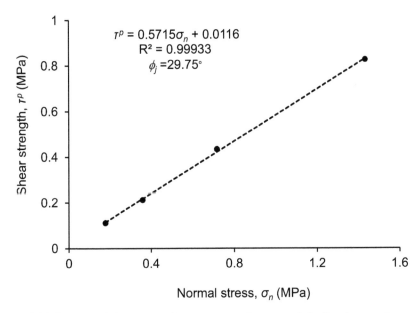

Figure A.66 Diagram of shear strength versus normal stress and the fitted regression curve for the interface between dacite and limestone (LDS-1)

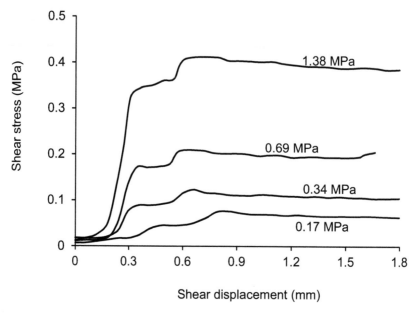

Figure A.67 Diagram of shear stress versus shear displacement curve for the interface between dacite and limestone (DLS-2)

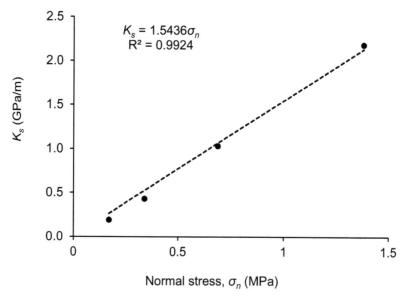

Figure A.68 Diagram of K_s versus normal stress and the fitted regression curve for the interface between dacite and limestone (DLS-2)

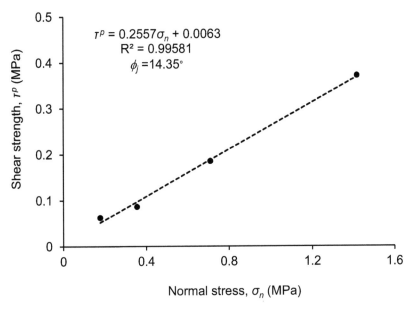

Figure A.69 Diagram of shear strength versus normal stress and the fitted regression curve for the interface between dacite and limestone (DLS-2)

Appendix B

Supplementary of the numerical modeling results

Table B.1 Maximum displacements around tunnels for Cases 3, 17 and 29–32

Case No.	Displacements around Tunnel 1 (cm)				Displacements around Tunnel 2 (cm)				Displacements around Tunnel 3 (cm)			
	Left wall	Right wall	Roof	Floor	Left wall	Right wall	Roof	Floor	Left wall	Right wall	Roof	Floor
3	−1.6	6.6	−7.9	5.8	−5.2	5.3	−4.9	4.1	4.7	−3.9	−4.2	4.3
17	−1.6	6.1	−7.4	5.7	−4.8	5.1	−4.8	4.0	4.5	−3.8	−4.0	4.2
29	−1.8	5.7	−6.6	4.4	−4.6	5.0	−4.6	3.5	4.2	−3.4	−3.7	3.4
30	−1.6	5.7	−6.4	4.4	−4.5	4.9	−4.4	3.5	4.1	−3.4	−3.7	3.4
31	−1.5	5.7	−6.2	4.0	−4.7	5.0	−4.5	3.5	4.1	−3.4	−3.7	3.4
32	−1.6	5.3	−5.8	3.6	−4.7	4.9	−4.4	3.5	3.9	−3.4	−3.6	3.2
33	−1.6	5.7	−6.3	4.2	−4.5	4.9	−4.4	3.5	4.1	−3.4	−3.6	3.4

Table B.2 Average failure zone thicknesses around tunnels for Cases 3, 17 and 29–32

Case No.	Failure zone thicknesses around Tunnel 1 (m)			Failure zone thicknesses around Tunnel 2 (m)				Failure zone thicknesses around Tunnel 3 (m)			
	Right wall	Roof	Floor	Left wall	Right wall	Roof	Floor	Left wall	Right wall	Roof	Floor
3	3.53	3.27	2.70	5.17	3.23	1.40	2.87	3.53	2.20	1.63	2.37
17	3.43	2.97	2.70	4.13	3.23	1.40	2.40	3.23	2.17	1.30	2.37
29	3.43	2.47	2.47	3.90	2.87	1.40	2.33	2.87	2.17	1.30	2.10
30	3.30	2.80	2.47	3.57	2.87	1.40	2.33	2.87	2.17	1.30	2.10
31	3.17	2.13	2.47	3.57	2.87	1.40	2.33	2.87	2.17	1.30	2.10
32	3.17	2.13	1.60	3.90	2.87	1.40	2.33	2.87	2.17	1.30	2.10
33	3.20	2.47	2.47	3.57	2.87	1.40	2.33	2.87	2.17	1.30	2.10

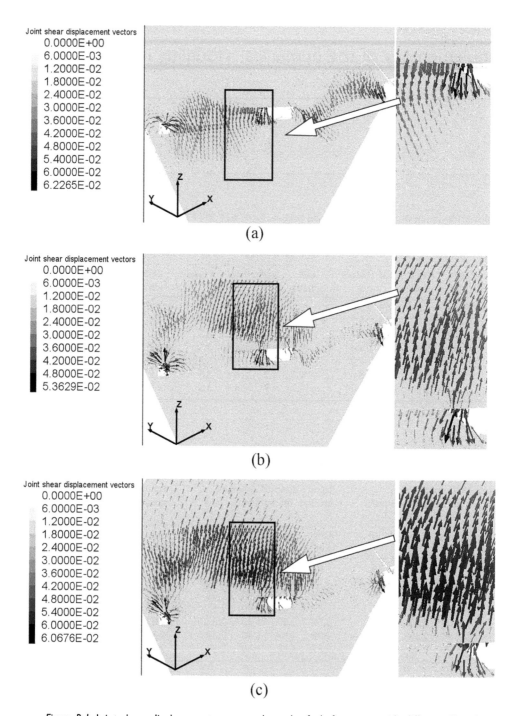

Figure B.1 Joint shear displacement vectors along the fault for cases with different K_0 values (unit: m) (observed from the hanging wall direction): (a) Case 2; (b) Case 4; (c) Case 5

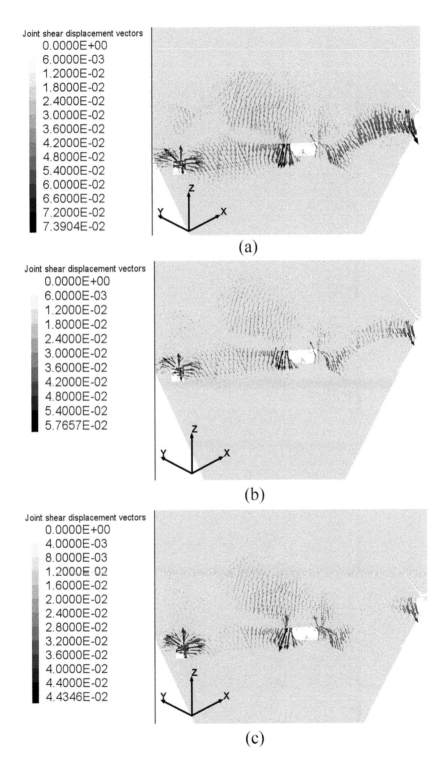

Figure B.2 Joint shear displacement vectors along the fault for different rock mass conditions (unit: m) (observed from the hanging wall direction): (a) Case 7; (b) Case 3; (c) Case 8

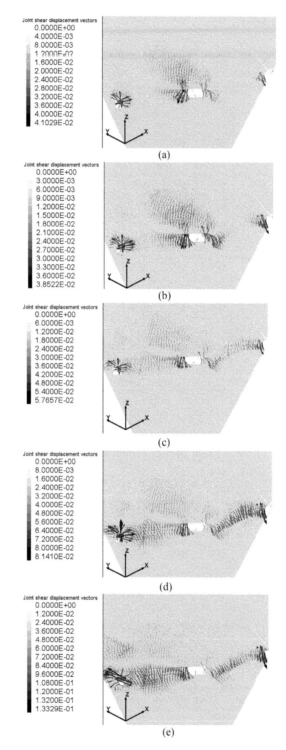

Figure B.3 Joint shear displacement vectors along the fault for cases with different post-failure residual strengths (unit: m) (observed from the hanging wall direction): (a) Case 9; (b) Case 10; (c) Case 3; (d) Case 11; (e) Case 12

References

Ahn, T. Y. & Song, J. J. (2011) New contact-definition algorithm using inscribed spheres for 3D discontinuous deformation analysis. *International Journal of Computational Methods*, 8(2), 171–191.

Alejano, L. R., Ramírez-Oyanguren, P. & Taboada, J. (1999) FDM predictive methodology of subsidence due to flat and inclined coal seam mining. *International Journal of Rock Mechanics Sciences and Geomechanics Abstracts*, 36(4), 475–491.

Alejano, L. R., Rodriguez-Dono, A., Alonso, E. & Fdez-Manín, G. (2009) Ground reaction curves for tunnels excavated in different quality rock masses showing several types of post-failure behavior. *Tunnelling and Underground Space Technology*, 24, 689–705.

Alejano, L. R., Alonso, E., Rodriguez-Dono, A. & Fernández-Manín, G. (2010) Application of the convergence-confinement method to tunnels in rock masses exhibiting Hoek–Brown strain-softening behavior. *International Journal of Rock Mechanics and Mining Sciences*, 47(1), 150–160.

Alejano, L. R., Alfonso, R. D. & Veiga, M. (2012) Plastic radii and longitudinal deformation profiles of tunnels excavated in strain-softening rock masses. *Tunnelling and Underground Space Technology*, 30, 169–182.

Amadei, B. & Goodman, R. E. (1981) A 3D constitutive relation for fractured rock masses. In: Selvadurai, A.P.S. (ed.) *Studies in Applied Mechanics, Part 5B, Mechanics of Structured Media: Proceedings of International Symposium on Mechanical Behavior of Structured Media, Ottawa*. Elsevier, New York. pp. 249–268.

ASTM D2845 (2008) *Standard Test Method for Laboratory Determination of Pulse Velocities and Ultrasonic Elastic Constants of Rock*. ASTM International, West Conshohocken, PA.

ASTM D3967 (2008) *Standard Test Method for Splitting Tensile Strength of Intact Rock Core Specimens*. ASTM International, West Conshohocken, PA.

ASTM D4543 (2008) *Standard Practices for Preparing Rock Core as Cylindrical Test Specimens and Verifying Conformance to Dimensional and Shape Tolerances*. ASTM International, West Conshohocken, PA.

ASTM D5607 (2008) *Standard Test Method for Performing Laboratory Direct Shear Strength Tests of Rock Specimens Under Constant Normal Force*. ASTM International, West Conshohocken, PA.

ASTM D7012 (2014) *Standard Test Methods for Compressive Strength and Elastic Moduli of Intact Rock Core Specimens under Varying States of Stress and Temperatures*. ASTM International, West Conshohocken, PA.

Atlas Copco. (2008) *Atlas Copco Rock Reinforcement Products: Product Catalogue*. Available from: https://www.corporacionfont.com/wp-content/uploads/2016/05/Sistema-de-anclaje.pdf.

Aydan, O., Ulusay, R. & Kawamoto, T. (1997) Assessment of rock mass strength for underground excavations. *International Journal of Rock Mechanics and Mining Sciences*, 34,18.e1–18.e17.

Bandis, S., Lumsden, A. C. & Barton, N. R. (1981) Experimental studies of scale effects on the shear behavior of rock joints. *International Journal of Rock Mechanics and Mining Sciences & Geomechanics Abstracts*, 18, 1–21.

Bandis, S., Lumsden, A. C. & Barton, N. R. (1983) Fundamentals of rock joint deformation. *International Journal of Rock Mechanics and Mining Sciences & Geomechanics Abstracts*, 20(6), 249–268.

Barla, G. (2001) Tunnelling under squeezing rock conditions. In: Kolymbas, D. (ed.) *Proceedings of Eurosummer-School in Tunnel Mechanics, Innsbruck*. Logos Verlag, Berlin. pp. 169–268.

Barla, G. & Barla, M. (2000) Continuum and discontinuum modeling in tunnel engineering. *Rudarsko-Geološko-Naftni Zbornik*, 12, 45–57.

Barla, M., Piovano, G. & Grasselli, G. (2011) Rock slide simulation with the combined finite discrete element method. *International Journal of Geomechanics*, 12(6), 711–721.

Barton, N. (1972) A model study of rock-joint deformation. *International Journal of Rock Mechanics and Mining Sciences*, 9, 579–602.

Barton, N. (2002) Some new Q-value correlations to assist in site characterization and tunnel design. *International Journal of Rock Mechanics and Mining Sciences*, 39, 185–216.

Barton, N. & Choubey, V. (1977) The shear strength of rock joints in theory and practice. *Rock Mechanics*, 10, 1–54.

Barton, N., Lien, R. & Lunde, J. (1974) Engineering classification of rock masses for the design of tunnel support. *Rock Mechanics*, 6(4), 189–236.

Barton, N., Bandis, S. & Bakhtar, K. (1985) Strength, deformation and conductivity coupling of rock joints. *International Journal of Rock Mechanics and Mining Sciences & Geomechanics Abstracts*, 22, 120–140.

Beer, G. & Poulsen, B. A. (1994) Efficient numerical modelling of faulted rock using the boundary element method. *International Journal of Rock Mechanics and Mining Sciences & Geomechanics Abstracts*, 31(5), 485–506.

Benedetto, M. F., Berrone, S., Pieraccini, S. & Scialo, S. (2014) The virtual element method for discrete fracture network simulations. *Computer Methods in Applied Mechanics and Engineering*, 280, 135–156.

Beyabanaki, S.A.R., Mikola, R. G. & Hatami, K. (2008) Three-dimensional discontinuous deformation analysis (3-D DDA) using a new contact resolution algorithm. *Computers and Geotechnics*, 35(3), 346–356.

Bhasin, R. & Grimstad, E. (1996) The use of stress-strength relationships in the assessment of tunnel stability. *Tunnelling and Underground Space Technology*, 11(1), 93–98.

Bhasin, R. & Hoeg, K. (1997) Parametric study for a large cavern in jointed rock using a distinct element model (UDEC-BB). *International Journal of Rock Mechanics and Mining Sciences*, 35(1), 17–29.

Bhasin, R., Magnussen, A. W. & Grimstad, E. (2006) The effect of tunnel size on stability problems in rock masses. *Tunnelling and Underground Space Technology*, 21, 405.

Bieniawski, Z. T. (1973) Engineering classification of jointed rock masses. *Transactions of the South African Institution of Civil Engineers*, 15, 335–344.

Bieniawski, Z. T. (1976) Rock mass classifications in rock engineering. In: Bieniawski, Z. T. (ed.) *Proceedings Symposium on Exploration for Rock Engineering, Johannesburg*. Balkema, Cape Town. 1. pp. 97–106.

Bieniawski, Z. T. (1978) Determining rock mass deformability: Experience from case histories. *International Journal of Rock Mechanics and Mining Sciences*, 15, 237–247.

Bieniawski, Z. T. (1988) The Rock Mass Rating (RMR) system (geomechanics classification) in engineering practice. In: L. Kirkaldie (ed.) *Proceedings of Rock Classification Systems for Engineering Purposes*. American Society for Testing and Materials, Philadelphia. pp. 17–34.

Bieniawski, Z. T. (1989) *Engineering Rock Mass Classifications: A complete manual for engineers and geologists in mining, civil, and petroleum engineering*. Wiley, New York.

Bieniawski, Z. T. (1993) Classification of rock masses for engineering: The RMR system and future trends. *Comprehensive Rock Engineering*, 3, 553–573.

Bieniawski, Z.T. & Van Heerden, W.L. (1975) The significance of in-situ tests on large rock speci-mens. *International Journal of Rock Mechanics and Mining Sciences*, 12, 101–113.

Blandford, G. E., Ingraffea, A. R. & Liggett, J. A. (1981) Two-dimensional stress intensity factor computations using boundary element method. *International Journal for Numerical Methods in Engineering*, 17, 387–404.

Bobet, A. & Einstein, H. H. (2011) Tunnel reinforcement with rockbolts. *Tunnelling and Under-ground Space Technology*, 26, 100–123.

Brady, B.H.G. & Bray, J. W. (1978) The boundary element method for determining stress and dis-placements around long openings in a triaxial stress field. *International Journal of Rock Mechan-ics and Mining Sciences & Geomechanics Abstracts*, 15, 21–28.

Brady, B.H.G. & Brown, E. T. (2004) *Rock Mechanics for Underground Mining*. Kluwer Academic, Dordrecht.

Brown, E. T. (1970) Strength of models of rock with intermittent joints. *Journal of the Soil Mechanics and Foundations Division, ASCE*, 96, 1935–1949.

Brown, E. T. & Hoek, E. (1978) Trends in relationships between measured in-situ stresses and depth. *International Journal of Rock Mechanics and Mining Sciences & Geomechanics Abstracts*, 15, 211–215.

Brown, E. T., Bray, J. W., Landayi, B. & Hoek, E. (1983) Ground response curves for rock tunnels. *Journal of Geotechnical Engineering Division, ASCE*, 109(1), 15–39.

Bruneau, G., Hudyma, M. R., Hadjigeorgiou, J. & Potvin, Y. (2003) Influence of faulting on a mine shaft: A case study: Part I: Background and instrumentation. *International Journal of Rock Mechanics and Mining Sciences*, 40, 95–111.

Bucky, P. B. (1934) Effect of approximately vertical cracks on the behaviour of horizontally lying roof strata. *Transaction of American Institute of Mining, Metallurgical, and Petroleum Engineers*, 109, 212–229.

Bucky, P. B. & Taborelli, R. V. (1938) Effects of immediate roof thickness in longwall mining as determined by barodynamic experiments. *Transaction of American Institute of Mining, Metallur-gical, and Petroleum Engineers*, 130, 314–332.

Cai, M. (2008) Influence of stress path on tunnel excavation response: Numerical tool selection and modeling strategy. *Tunnelling and Underground Space Technology*, 23, 618–628.

Cai, M. & Kaiser, P. K. (2004) Numerical simulation of the Brazilian test and the tensile strength of anisotropic rocks and rocks with pre-existing cracks. *International Journal of Rock Mechanics and Mining Sciences*, 41(Suppl. 1), 478–483.

Cai, M., Kaiser, P. K., Tasaka, Y. & Minami, M. (2007a) Determination of residual strength param-eters of jointed rock masses using the GSI system. *International Journal of Rock Mechanics and Mining Sciences*, 44(2), 247–265.

Cai, M., Morioka, H., Kaiser, P. K., Tasaka, Y., Minami, M. & Maejima, T. (2007b) Back analysis of rock mass strength parameters using AE monitoring data. *International Journal of Rock Mechan-ics and Mining Sciences*, 44(4), 538–549.

Cantieni, L. & Anagnostou, G. (2009) The effect of the stress path on squeezing behavior in tunnel-ing. *Rock Mechanics and Rock Engineering*, 42, 289–318.

Carranza-Torres, C. & Fairhurst, C. (1999) The elasto-plastic response of underground excavations in rock masses that satisfy the Hoek-Brown failure criterion. *International Journal of Rock Mechan-ics and Mining Sciences*, 36(6), 777–809.

Carranza-Torres, C. & Fairhurst, C. (2000) Application of convergence-confinement method of tunnel design to rock masses that satisfy the Hoek–Brown failure criterion. *Tunnelling and Underground Space Technology*, 15(2), 187–213.

Cecil, O. S. (1970) *Correlations of Rock Bolt-Shotcrete Support and Rock Quality Parameters in Scandinavian Tunnels*. Ph.D. Thesis, University of Illinois, Urbana.

Chen, C. S., Pan, E. & Amadei, B. (1998) Fracture mechanics analysis of cracked discs of anisotropic rock using the boundary element method. *International Journal of Rock Mechanics and Mining Sciences*, 35(2), 195–218.

Chen, H. M., Zhao, Z. Y., Choo, L. Q. & Sun, J. P. (2016) Rock cavern stability analysis under different hydro-geological conditions using the coupled hydro-mechanical model. *Rock Mechanics and Rock Engineering*, 49, 555–572.

Chen, S. G. & Zhao, J. (2002) Modeling of tunnel excavation using a hybrid DEM/BEM method. *Computer-Aided Civil and Infrastructure Engineering*, 17, 381–386.

Chryssanthakis, P., Barton, N., Lorig, L. & Christianson, M. (1997) Numerical simulation of fiber reinforced shotcrete in tunnel using the discrete element method. *International Journal of Rock Mechanics and Mining Sciences*, 34, 54.e1–54.e14.

Chu, B. L., Hsu, S. C., Chang, Y. L. & Lin, Y. S. (2007) Mechanical behavior of a twin-tunnel in multi-layered formations. *Tunnelling and Underground Space Technology*, 22(3), 351–362.

Coggan, J., Gao, F., Stead, D. & Elmo, D. (2012) Numerical modelling of the effects of weak immediate roof lithology on coal mine roadway stability. *International Journal of Coal Geology*, 90, 100–109.

Cording, E. J. & Deere, D. U. (1972) Rock tunnel supports and field measurements. In: Lane, K. S. & Garfield, L. A. (eds.) *Proceedings of North American Rapid Excavation and Tunneling Conference, Chicago*. American Institute of Mining, Metallurgical, and Petroleum Engineers, New York. 1. pp. 567–600.

Crouch, S. L. & Starfield, A. M. (1983) *Boundary element methods in solid mechanics*. George Allen & Unwin, London.

Crowder, J. J. & Bawden, W. F. (2006) *The Estimation of Post-Peak Rock Mass Properties: Numerical Back Analysis Calibrated Using in Situ Instrumentation Data*. Rocnews. Available from: http://www.rocscience.com/library/rocnews.

Cui, K., Défossez, P. & Richard, G. (2007) A new approach for modelling vertical stress distribution at the soil/tyre interface to predict the compaction of cultivated soils by using the PLAXIS code. *Soil & Tillage Research*, 95, 277–287.

Cui, Z., Sheng, Q. & Leng, X. (2016) Control effect of a large geological discontinuity on the seismic response and stability of underground rock caverns: A case study of Baihetan #1 surge chamber. *Rock Mechanics and Rock Engineering*, 49, 2099–2114.

Cundall, P. A. (1971) A computer model for simulating progressive, large-scale movements in blocky rock systems. *Rock Fracture: Proceedings of the International Symposium Rock Fracture, Nancy, France*. ISRM. 2. pp. 129–136.

Cundall, P. A. (1980) *UDEC-A Generalized Distinct Element Program for Modelling Jointed Rock*. Report from P. Cundall Associates to U.S. Army European Research Office, London.

Cundall, P. A. (1988) Formulation of a three-dimensional distinct element model, Part I: A scheme to detect and represent contacts in a system composed of many polyhedral blocks. *International Journal of Rock Mechanics and Mining Sciences & Geomechanics Abstracts*, 25, 107–116.

Cundall, P. A. (2001) A discontinuous future for numerical modelling in geomechanics? *Proceedings of the Institution of Civil Engineering-Geotechnical Engineering*, 149(1), 41–47.

Deere, D. U. (1968) Geological considerations. In: Stagg, K. G. & Zienkiewicz, O. C. (eds.) *Proceedings of Rock Mechanics in Engineering Practice*. Wiley, London. pp. 1–20.

Deere, D. U. & Deere, D. W. (1988) The rock quality designation (RQD) index in practice. In: Kirkaldie, L. (ed.) *Proceedings of Symposium on Rock Classification Systems for Engineering Purposes*. ASTM Special Technical Publication 984, Philadelphia. pp. 91–101.

Deere, D. U., Hendron, A. J., Patton, F. D. & Cording, E. J. (1967) Design of surface and near surface construction in rock. In: Fairhurst, C. (ed.) *Failure and Breakage of Rock: Proceedings of the 8th US Symposium on Rock Mechanics*. American Institute of Mining, Metallurgical, and Petroleum Engineers, New York. pp. 237–302.

Deere, D. U., Peck, R. B., Monsees, J. E., Parker, H. W., Monsees, J. E. & Schmidt, B. (1970) Design of tunnel support system. *Highway Research Record*, 339, 12–16.

Diederichs, M. S. & Kaiser, P. K. (1999) Stability of large excavations in laminated hard rock masses: The voussoir analogue revisited. *International Journal of Rock Mechanics and Mining Sciences*, 36, 97–117.

Eberhardsteiner, J., Mang, H. A. & Torzicky, P. (1993) Hybrid BE-FE stress analysis of the excavation of a tunnel bifurcation on the basis of a substructuring technique. *Advances in Boundary Element Techniques*, 103–128.

Eberhardt, E. (2001) Numerical modeling of three-dimensional stress rotation ahead of an advancing tunnel face. *International Journal of Rock Mechanics and Mining Sciences*, 38(4), 499–518.

Eberhardt, E., Stead, D. & Coggan, J. S. (2004) Numerical analysis of initiation and progressive failure in natural rock slopes: The 1991 Randa rockslide. *International Journal of Rock Mechanics and Mining Sciences*, 41(1), 69–87.

Egger, P. (2000) Design and construction aspects of deep tunnels (with particular emphasis on strain softening rocks). *Tunnelling and Underground Space Technology*, 15(4), 403–408.

Elmo, D. (2006) *Evaluation of a Hybrid FEM/DEM Approach for Determination of Rock Mass Strength Using a Combination of Discontinuity Mapping and Fracture Mechanics Modelling, with Particular Emphasis on Modelling of Jointed Pillars*. Ph.D. Thesis, University of Exeter, Exeter, UK.

Elmo, D. & Stead, D. (2010) An integrated numerical modelling-discrete fracture network approach applied to the characterisation of rock mass strength of naturally fractured pillars. *Rock Mechanics and Rock Engineering*, 43(1), 3–19.

Evans, W. H. (1941) The strength of undermined strata. *Transaction of The Institution of Mining and Metallurgy*, 50, 475–500.

Fahimifar, A. & Reza Zareifard, M. (2009) A theoretical solution for analysis of tunnels below groundwater considering the hydraulic-mechanical coupling. *Tunnelling and Underground Space Technology*, 24(6), 634–646.

Fahimifar, A. & Soroush, H. (2005) A theoretical approach for analysis of the interaction between grouted rockbolts and rock masses. *Tunnelling and Underground Space Technology*, 20, 333–343.

Falls, S. D. & Young, R. P. (1998) Acoustic emission and ultrasonic-velocity methods used to characterize the excavation disturbance associated with deep tunnels in hard rock. *Tectonophysics*, 289, 1–15.

Fayol, M. (1885) Sur les movements de terrain provoques par l'exploitation des mines. *Bulletin Society of Industries and Mining*, 14, 818.

Fekete, S. & Diederichs, M. (2013) Integration of three-dimensional laser scanning with discontinuum modelling for stability analysis of tunnels in blocky rock masses. *International Journal of Rock Mechanics and Mining Sciences*, 57, 11–23.

Feng, J., Wang, G. & Zhang, C. (1998) A hybrid procedure of distinct: Boundary element for discrete rock dynamic analysis. *Developments in Geotechnical Engineering*, 83, 313–320.

Feng, Z. L. & Lewis, R. W. (1987) Optimal estimation of in-situ ground stress from displacement measurements. *International Journal for Numerical and Analytical Methods in Geomechanics*, 11, 397–408.

Fossum, A. F. (1985) Effective elastic properties for a randomly jointed rock mass. *International Journal of Rock Mechanics and Mining Sciences & Geomechanics Abstracts*, 22, 467–470.

Gao, F., Stead, D. & Kang, H. (2014) Simulation of roof shear failure in coal mine roadways using an innovative UDEC Trigon approach. *Computers and Geotechnics*, 61, 33–41.

Gens, A., Carol, I. & Alonso, E. E. (1995) Rock joints: FEM implementation and applications. In: Selvadurai, A.P.S. & Boulon, M. J. (eds.) *Mechanics of Geomaterial Interfaces*. Elsevier, Amsterdam. pp. 395–420.

Gerrard, C. M. (1982) Elastic moduli of rock masses having one, two and three sets of joints. *International Journal of Rock Mechanics and Mining Sciences & Geomechanics Abstracts*, 19, 15–23.

Ghaboussi, J., Wilson, E. L. & Isenberg, J. (1973) Finite element for rock joints and interfaces. *Journal of the Soil Mechanics and Foundations Division, ASCE*, 96, SM10, 833–848.

Gonzalez de Vallejo, L. I. (1983) A new rock classification system for underground assessment using surface data. *Proceedings of the International Symposium on Engineering Geology and Underground Construction, LNEC, Lisbon, Portugal*. Balkema, Rotterdam. pp. 85–94.

Goodman, R. E. (1974) The mechanical properties of joints. *Proceedings of the Third International Congress of the International Society of Rock Mechanics, Denver, Colorado*. National Academy of Sciences, Washington, DC. 1. pp. 127–140.

Goodman, R. E. (1989) *Introduction to Rock Mechanics*. John Wiley and Sons, New York.

Goodman, R. E., Taylor, R. L. & Brekke, T. L. (1968) A model for the mechanics of jointed rock. *Journal of the Soil Mechanics and Foundations Division, ASCE*, 94, SM3, 637–659.

Grimstad, E. & Barton, N. (1993) Updating the Q-system for NMT. In: Kompen, C., Opsahl, S. L. & Berg, S. L. (eds.) *Proceedings of the International Symposium on Sprayed Concrete-Modern Use of Wet Mix Sprayed Concrete for Underground Support, Fagernes, Norway*. Norwegian Concrete Association, Oslo. pp. 20.

Guan, Z., Jiang, Y. & Tanabasi, Y. (2007) Ground reaction analyses in conventional tunnelling excavation. *Tunnelling and Underground Space Technology*, 22(2), 230–237.

Hajiabdolmajid, V., Kaiser, P. K. & Martin, C. D. (2002) Modelling brittle failure of rock. *International Journal of Rock Mechanics and Mining Sciences*, 39, 731–741.

Hao, Y. H. & Azzam, R. (2005) The plastic zones and displacements around underground openings in rock masses containing a fault. *Tunnelling and Underground Space Technology*, 20, 49–61.

Hart, R., Cundall, P. A. & Lemos, J. (1988) Formulation of a three-dimensional distinct element model, Part II: Mechanical calculations for motion and interaction of a system composed of many polyhedral blocks. *International Journal of Rock Mechanics and Mining Sciences & Geomechanics Abstracts*, 25(3), 117–125.

Hoek, E. (1994) Strength of rock and rock masses. *ISRM New Journal*, 2(2), 4–16.

Hoek, E. & Brown, E. T. (1980) *Underground Excavations in Rock*. The Institute of Mining and Metallurgy, London.

Hoek, E. & Brown, E. T. (1988) The Hoek-Brown failure criterion-a 1988 update. In: Curran, J. C. (ed.) *Proceedings of 15th Canadian Rock Mechanics Symposium, Toronto*. Univ. Toronto Press, Toronto. pp. 31–38.

Hoek, E. & Brown, E. T. (1997) Practical estimates of rock mass strength. *International Journal of Rock Mechanics and Mining Sciences*, 34, 1165–1186.

Hoek, E. & Diederichs, M. S. (2006) Empirical estimation of rock mass modulus. *International Journal of Rock Mechanics and Mining Sciences*, 43, 203–215.

Hoek, E., Kaiser, P. K. & Bawden, W. F. (1995) *Support of Underground Excavations in Hard Rock*. Balkema, Rotterdam.

Hoek, E., Carranza-Torres, C. & Corkum, B. (2002) Hoek-Brown criterion-2002 edition. In: Hammah, R., Bawden, W., Curran, J. & Telesnicki, M. (eds.) *Proceedings of the Fifth North American Rock Mechanics Symposium, Toronto*. Univ. Toronto Press, Toronto. 1. pp. 267–273.

Hsiao, F. Y., Wang, C. L. & Chern, J. C. (2009) Numerical simulation of rock deformation for support design in tunnel intersection area. *Tunnelling and Underground Space Technology*, 24, 14–21.

Hu, K. X. & Huang, Y. (1993) Estimation of the elastic properties of fractured rock masses. *International Journal of Rock Mechanics and Mining Sciences & Geomechanics Abstracts*, 30(4), 381–394.

Huang, F., Zhu, H., Xu, Q., Cai, Y. & Zhuang, X. (2013) The effect of weak interlayer on the failure pattern of rock mass around tunnel-scaled model tests and numerical analysis. *Tunnelling and Underground Space Technology*, 35(8), 207–218.

Huang, G., Kulatilake, P.H.S.W., Cai, S. & Song, H. (2017) 3-D discontinuum numerical modeling of subsidence due to ore extraction and backfilling operations in an underground iron mine in China. *International Journal of Mining Science and Technology*, 27, 191–201.

Hudson, J. A. (2001) Rock engineering case histories: Key factors, mechanisms and problems. In: Elorante, P. & Sarkka, P. (eds.) *Rock Mechanics: A Challenge for Society: Proceedings of the ISRM Regional Symposium Eurock, Espoo, Finland*. Balkema, Rotterdam. pp. 13–20.

Itasca Consulting Group, Inc. (2007) 3DEC-3 dimensional distinct element code, version 4.1.

Itasca Consulting Group, Ltd. (1993) FLAC manuals.

Janin, J. P., Dias, D., Emeriault, F., Kastner, R., Bissonnais, H. L. & Guilloux, A. (2015) Numerical back-analysis of the southern Toulon tunnel measurements: A comparison of 3D and 2D approaches. *Engineering Geology*, 195, 42–52.

Jia, P. & Tang, C. A. (2008) Numerical study on failure mechanism of tunnel in jointed rock mass. *Tunnelling and Underground Space Technology*, 23, 500–507.

Jiang, L., Mitri, H. S., Ma, N. & Zhao, X. (2016) Effect of foundation rigidity on stratified roadway roof stability in underground coal mines. *Arabian Journal of Geosciences*, 9, 32.

Jiang, Q. H. & Yeung, M. R. (2004) A model of point-to-face contact for three-dimensional discontinuous deformation analysis. *Rock Mechanics and Rock Engineering*, 37(2), 95–116.

Jing, L. (2003) A review of techniques, advances and outstanding issues in numerical modeling for rock mechanics and rock engineering. *International Journal of Rock Mechanics and Mining Sciences*, 40, 283–353.

Jing, L. & Hudson, J. A. (2002) Numerical methods in rock mechanics. *International Journal of Rock Mechanics and Mining Sciences*, 39, 409–427.

Kaiser, P. K., Zou, D. & Lang, P. A. (1990) Stress determination by back-analysis of excavation-induced stress changes: A case study. *Rock Mechanics and Rock Engineering*, 23(3), 185–200.

Kaiser, P. K., Yazici, S. & Maloney, S. (2001) Mining induced stress change and consequences of stress path on excavation stability: A case study. *International Journal of Rock Mechanics and Mining Sciences*, 38(2), 167–180.

Kalamaris, G. S. & Bieniawski, Z. T. (1995) A rock mass strength concept for coal incorporating the effect of time. In: Fuji, T. (ed.) *Proceedings of the Eighth International Congress on Rock Mechanics, Tokyo*. Balkema. Rotterdam. 1. pp. 295–302.

Karami, A. & Stead, D. (2008) Asperity degradation and damage in the direct shear test: A hybrid FEM/DEM approach. *Rock Mechanics and Rock Engineering*, 41(2), 229–266.

Kavvadas, M. (2005) Monitoring ground deformation in tunnelling: Current practice in transportation tunnels. *Engineering Geology*, 79, 89–109.

Khan, M. S. (2010) *Investigation of Discontinuous Deformation Analysis for Application in Jointed Rock Masses*. Ph.D. Thesis, University of Toronto.

Kim, Y., Amadei, B. & Pan, E. (1999) Modeling the effect of water, excavation sequence and rock reinforcement with discontinuous deformation analysis. *International Journal of Rock Mechanics and Mining Sciences*, 36, 949–970.

Kimura, F., Okabayashi, N. & Kawamoto, T. (1987) Tunnelling through squeezing rock in two large fault zones of the Enasan Tunnel II. *Rock Mechanics and Rock Engineering*, 20, 151–166.

Kirsch, G. (1898) Die Theorie der Elastizitat und die Beaurforisse der Festigkeitslehre. *Zeitschrift des Vereines deutscher Ingenieure*, 42, 797–807.

Klerck, P. A. (2000) *The Finite Element Modelling of Discrete Fracture of Brittle Materials*. Ph.D. Thesis, University of Wales, Swansea.

Klerck, P. A., Sellers, E. J. & Owen, D.R.J. (2004) Discrete fracture in quasi-brittle materials under compressive and tensile stress states. *Computer Methods in Applied Mechanics and Engineering*, 193, 3035–3056.

Kovari, K. & Amstad, CH. (1979) Field instrumentation in tunnelling as a practical design aid. *Proceedings of the 4th International Conference for Rock Mechanics, 2–8 September, Montreux, Switzerland*. ISRM. 2. pp. 311–318.

Kristen, H.A.D. (1976) Determination of rock mass elastic moduli by back analysis of deformation measurements. In: Bieniawaski, Z.T. (ed.) *Proceedings of Symposium on Exploration for Rock Engineering, Johannesburg*. Balkema, Cape Town. pp. 1154–1160.

Kulatilake, P.H.S.W. (1985) Estimating elastic constants and strength of discontinuous rock. *Journal of Geotechnical Engineering, ASCE,* 111, 847–864.

Kulatilake, P.H.S.W. (1998) Software manual for FRACNTWK: A computer package to model discontinuity geometry in rock masses. Technical report submitted to Metropolitan Water District of Southern California (MWD).

Kulatilake, P.H.S.W. & Shu, B. (2015) Prediction of rock mass deformations in three dimensions for a part of an open pit mine and comparison with field deformation monitoring data. *Geotechnical and Geological Engineering,* 33, 1551–1568.

Kulatilake, P.H.S.W., Wathugala, D. N. & Stephansson, O. (1993a) Joint network modelling, including a validation to an area in Stripa mine, Sweden. *International Journal of Rock Mechanics and Mining Sciences,* 30(5), 503–526.

Kulatilake, P.H.S.W., Wang, S. & Stephansson, O. (1993b) Effect of finite size joints on deformability of jointed rock at the three-dimensional level. *International Journal of Rock Mechanics and Mining Sciences,* 30(5), 479–501.

Kulatilake, P.H.S.W., Ucpirti, H., Wang, S. & Stephansson, O. (1992) Use of the distinct element method to perform stress analysis in rock with non-persistent joints and to study the effect of joint geometry parameters on the strength and deformability of rock masses. *Rock Mechanics Rock Engineering,* 25, 253–274.

Kulatilake, P.H.S.W., Chen, J., Teng, J., Shufang, X. & Pan, G. (1996) Discontinuity geometry characterization for the rock mass around a tunnel close to the permanent shiplock area of the three Gorges Dam Site in China. *International Journal of Rock Mechanics and Mining Sciences,* 33, 255–277.

Kulatilake, P.H.S.W., He, W., Um, J. & Wang, H. (1997) A physical model study of jointed rock mass strength under uniaxial compressive loading. *International Journal of Rock Mechanics and Mining Sciences,* 34, 623–633.

Kulatilake, P.H.S.W., Um, J., Wang, M., Escandon, R. F. & Varvaiz, J. (2003) Stochastic fracture geometry modeling in 3-D including validations for a part of Arrowhead East Tunnel, California, USA. *Engineering Geology,* 70, 131–155.

Kulatilake, P.H.S.W., Park, J. Y. & Um, J. G. (2004) Estimation of rock mass strength and deformability in 3-D for a 30 m cube at a depth of 485 m at Äspö Hard Rock Laboratory. *Geotechnical and Geological Engineering,* 22, 313–330.

Kulatilake, P.H.S.W., Shreedharan, S., Sherizadeh, T., Shu, B., Xing, Y. & He, P. (2016) Laboratory estimation of rock joint stiffness and frictional parameters. *Geotechnical and Geological Engineering,* 34(6), 1723–1735.

Lander, D. (2014) Advances in understanding stratigraphy and structure, Turquoise Ridge JV Mine. (Internal Presentation).

Lang, P. A., Chan, T., Davison, C. C., Everitt, R. A., Kozak, E. T. & Thompson, P. M. (1987) Near-field mechanical and hydraulic response of a granitic rock mass to shaft excavation. *Proceedings of the 28th U.S. Symposium on Rock Mechanics, 29 June–1 July, Tucson, Arizona.* American Rock Mechanics Association. pp. 509–516.

Laubscher, D. H. (1977) Geomechanics classification of jointed rock masses in mining applications. *Transaction of the Institution of Mining and Metallurgy,* A1–A8.

Laubscher, D. H. (1984) Design aspects and effectiveness of support systems in different mining conditions. *Transaction of the Institution of Mining and Metallurgy,* 93, A70–A82.

Lauffer, H. (1958) Gebirgsklassifizierung für den Stollenbau. *Geologie und Bauwesen,* 24(1), 46–51.

Lee, C. I. & Song, J. J. (2003) Rock engineering in underground energy storage in Korea. *Tunnelling and Underground Space Technology,* 18(5), 467–483.

Lorig, L. J., Brady, B.H.G. & Cundall, P. A. (1986) Hybrid distinct element boundary element analysis of jointed rock. *International Journal of Rock Mechanics and Mining Sciences & Geomechanics Abstracts,* 23(4), 303–312.

Lu, A., Xu, G., Sun, F. & Sun, W. (2010) Elasto-plastic analysis of a circular tunnel including the effects of the axial in situ stress. *International Journal of Rock Mechanics and Mining Sciences*, 47(1), 50–59.

Mahabadi, O.K., Cottrell, B. & Grasselli, G. (2010) An example of realistic modeling of rock dynamics problems: FEM/DEM simulation of dynamic Brazilian test on Barre granite. *Rock Mechanics and Rock Engineering*, 43(6), 707–716.

Mahabadi, O. K., Lisjak, A., Munjiza, A. & Grasselli, G. (2012) Y-Geo: New combined finite-discrete element numerical code for geomechanical applications. *International Journal of Geomechanics*, 12, 676–688.

Malama, B. & Kulatilake, P.H.S.W. (2003) Models for normal fracture deformation under compressive loading. *International Journal of Rock Mechanics and Mining Sciences*, 40, 893–901.

Marinos, P. & Hoek, E. (2000) GSI: A geologically friendly tool for rock mass strength estimation. *Proceedings of the GeoEng2000 at the International Conference on Geotechnical and Geological Engineering, Melbourne*. Technomic Publishers, Lancaster. pp. 1422–1446.

Marinos, P. & Hoek, E. (2001) Estimating the geotechnical properties of heterogeneous rock masses such as flysch. *Bulletin of Engineering Geology and Environment*, 60, 82–92.

Marti, J. & Cundall, P. (1982) Mixed discretization procedure for accurate modelling of plastic collapse. *International Journal for Numerical and Analytical Methods in Geomechanics*, 6(1), 129–139.

Martin, C. D. & Simmons, G. R. (1993) The atomic energy of Canada limited underground research laboratory: An overview of geomechanics characterization. In: Hudson, J. A. (ed.) *Comprehensive Rock Engineering: Principals, Practice & Projects*. Pergamon Press, Oxford. pp. 915–950.

Martin, C. D., Kaiser, P. K. & McCreath, D. R. (1999) Hoek–Brown parameters for predicting the depth of brittle failure around tunnels. *Canadian Geotechnical Journal*, 36(1), 136–151.

Martino, J. B. & Chandler, N. A. (2004) Excavation-induced damage studies at the underground research laboratory. *International Journal of Rock Mechanics and Mining Sciences*, 41, 1413–1426.

Maxwell, S. C., Young, R. P. & Read, R. S. (1998) A micro-velocity tool to assess the excavation damaged zone. *International Journal of Rock Mechanics and Mining Sciences*, 35(2), 235–247.

Merritt, A. H. (1972) Geologic prediction for underground excavation. In: Lane, K. S. & Garfield, L. A. (eds.) *Proceedings of the First Rapid Excavation and Tunneling Conference, Chicago*. American Institute of Mining, Metallurgical, and Petroleum Engineers, New York. pp. 115–132.

Mortazavi, A., Hassani, F. P. & Shabani, M. (2009) A numerical investigation of rock pillar failure mechanism in underground openings. *Computers and Geotechnics*, 36, 691–697.

Munjiza, A., Owen, D.R.J. & Bicanic, N. (1995) A combined finite-discrete element method in transient dynamics of fracturing solids. *Engineering Computations*, 12(2), 145–174.

Munjiza, A., Andrews, K. & White, J. (1999) Combined single and smeared crack model in combined finite-discrete element analysis. *International Journal for Numerical Methods in Engineering*, 44, 41–57.

Nagtegaal, J.C., Parks, D.M. & Rice, J.R. (1974) On numerically accurate finite element solutions in the fully plastic range. *Computer Methods in Applied Mechanics and Engineering*, 4(2), 153–177.

Ohnishi, Y., Sasaki, T., Koyama, T., Hagiwara, I., Miki, S. & Shimauchi, T. (2014) Recent insights into analytical precision and modelling of DDA and NMM for practical problems. *Geomechanics and Geoengineering*, 9(2), 97–11.

Pacher, F., Rabcewicz, L. & Golser, J. (1974) Zum der seitigen Stand der Gebirgs-klassifizierung in Stollen-und Tunnelbau. *Proceedings of XXII Geomechanical Colloquium, Salzburg*. Ministry of Building and Technology, Austria. pp. 51–58.

Palmström, A. (1982) The volumetric joint count-a useful and simple measure of the degree of rock jointing. *Proceedings of the Fourth International Congress on International Association on Engineering Geology, India*. Balkema, Rotterdam. pp. 221–228.

Palmström, A. & Stille, H. (2007) Ground behavior and rock engineering tools for underground excavations. *Tunnelling and Underground Space Technology*, 20, 362–377.

Pan, E., Amadei, B. & Kim, Y. I. (1998) 2D BEM analysis of anisotropic halfplane problems: Application to rock mechanics. *International Journal of Rock Mechanics and Mining Sciences*, 35(1),
69–74.

Please, C. P., Mason, D. P., Khalique, C. M., Ngnotchouye, J.M.T., Hutchinson, A. J., Van der Merwe,
J. N. & Yilmaz, H. (2013) Fracturing of an Euler–Bernoulli beam in coal mine pillar extraction.
International Journal of Rock Mechanics and Mining Sciences, 64, 132–138.

Prudencio, M. & Van Sint Jan, M. (2007) Strength and failure modes of rock mass models with non-
persistent joints. *International Journal of Rock Mechanics and Mining Sciences*, 44(6), 890–902.

Ray, A. K. (2009) *Influence of Cutting Sequence and Time Effects on Cutters and Roof Falls in
Underground Coal Mine-Numerical Approach*. Ph.D. Thesis, West Virginia University.

Read, R. S. (2004) 20 years of excavation response studies at AECL's Underground Research Laboratory. *International Journal of Rock Mechanics and Mining Sciences*, 41, 1251–1275.

Read, R. S. & Martin, C. D. (1992) Monitoring the excavation-induced response of granite. In: Til-
lerson, J. & Wawersik, W. (eds.) *Proceedings of the 33rd U.S. Symposium on Rock Mechanics,
3–5 June, Santa Fe, New Mexico*. Balkema, Rotterdam. pp. 201–210.

Riedmüller, G. & Schubert, W. (1999) Critical comments on quantitative rock mass classifications.
Felsbau, 17(3), 164–167.

Rockfield Software Ltd (2011) *ELFEN User Manual*. Swansea, UK.

Rocscience Inc. (2004) Phase 2, 5. Rocscience Inc. Toronto. Available from www.rocscience.com.

Romana, M. (1985) New adjustment ratings for application of Bieniawski classification to slopes.
Proceedings of International Symposium on the Role of Rock Mechanics, Zacatecas, Mexico.
ISRM. pp. 49–53.

Rose, D. (1982) Revising Terzaghi's tunnel rock load coefficients. *Proceedings of the 23rd US Symposium on Rock Mechanics*. American Institute of Mining, Metallurgical, and Petroleum Engineers. New York. pp. 953–960.

Sainsbury, B. L. & Sainsbury, D. P. (2017) Practical use of the ubiquitous-joint constitutive model
for the simulation of anisotropic rock masses. *Rock Mechanics and Rock Engineering*, 6(50),
1507–1528.

Sakurai, S. (1997) Lessons learned from field measurements in tunnelling. *Tunnelling and Underground Space Technology*, 12(4), 453–460.

Sakurai, S. & Takeuchi, K. (1983) Back analysis of measured displacements of tunnels. *Rock
Mechanics and Rock Engineering*, 16, 173–180.

Sandbak, L. A. & Rai, A. R. (2013) Ground support strategies at the Turquoise Ridge Joint Venture,
Nevada. *Rock Mechanics and Rock Engineering*, 46, 437–454.

Satyanarayana, I., Budi, G. & Deb, D. (2015) Strata behaviour during depillaring in Blasting Gallery
panel by field instrumentation and numerical study. *Arabian Journal of Geosciences*, 8, 6931–6947.

Sazid, M. & Singh, T. N. (2013) Two-dimensional dynamic finite element simulation of rock blasting.
Arabian Journal of Geosciences, 6(10), 3703–3708.

Sellers, E. J. & Klerck, P. A. (2000) Modelling of the effect of discontinuities on the extent of the
fracture zone surrounding deep tunnels. *Tunnelling and Underground Space Technology*, 15(4),
463–469.

Serafim, J. L. & Pereira, J. P. (1983) Considerations of the geomechanics classification of Bieniawski.
*Proceedings of International Symposium Engineering Geology and Underground Construction,
Lisbon*. Balkema, Rotterdam. 1. pp. 33–42.

Sharan, S. K. (2003) Elasto-brittle-plastic analysis of circular opening in Hoek–Brown media. *International Journal of Rock Mechanics and Mining Sciences*, 40, 817–824.

Sharan, S. K. (2005) Exact and approximate solutions for displacements around circular openings in
elastic-brittle-plastic Hoek-Brown rock. *International Journal of Rock Mechanics and Mining Sciences*, 42, 542–549.

Shen, B. & Barton, N. (1997) The disturbed zone around tunnels in jointed rock masses. *International Journal of Rock Mechanics and Mining Sciences*, 34(1), 117–125.

Sheorey, P. R. (1994) A theory for in situ stresses in isotropic and transversely isotropic rock. *International Journal of Rock Mechanics and Mining Sciences & Geomechanics Abstracts*, 31(1), 23–34.

Sherizadeh, T. & Kulatilake, P.H.S.W. (2016) Assessment of roof stability in a room and pillar coal mine in the U.S. using three-dimensional distinct element method. *Tunneling and Underground Space Technology*, 59, 24–37.

Shi, G. H. (1988) *Discontinuous Deformation Analysis: A New Numerical Model for the Statics and Dynamics of Block Systems*. Ph.D. Thesis, University of California, Berkeley.

Shi, G. H. (2014) Application of discontinuous deformation analysis on stability analysis of slopes and underground power houses. *Geomechanics and Geoengineering*, 9(2), 80–96.

Shi, G. H. & Goodman, R. E. (1985) Two-dimensional discontinuous deformation analysis. *International Journal for Numerical and Analytical Methods in Geomechanics*, 9(6), 541–556.

Shreedharan, S. & Kulatilake, P.H.S.W. (2016) Discontinuum-equivalent continuum analysis of the stability of the stability of tunnels in a deep coal mine using the distinct element method. *Rock Mechanics and Rock Engineering*, 49, 1903–1922.

Singh, B. (1973) Continuum characterization of jointed rock masses, part 1 and part 2. *International Journal of Rock Mechanics and Mining Sciences & Geomechanics Abstracts*, 10, 311–343.

Sitharam, T. G., Sridevi, J. & Shimizu, N. (2001) Practical equivalent continuum characterization of jointed rock masses. *International Journal of Rock Mechanics and Mining Sciences*, 38(3), 437–448.

Sofianos, A. I. (1996) Analysis and design of an underground hard rock voussoir beam roof. *International Journal of Rock Mechanics and Mining Sciences & Geomechanics Abstracts*, 33(2), 153–166.

Sonmez, H., Gokceoglu, C. & Ulusay, R. (2004) Indirect determination of the modulus of deformation of rock masses based on the GSI system. *International Journal of Rock Mechanics and Mining Sciences*, 41, 849–857.

Souley, M., Homand, F. & Thoraval, A. (1997) The effect of joint constitutive laws on the modelling of an underground excavation and comparison with in situ measurements. *International Journal of Rock Mechanics and Mining Sciences*, 34, 97–115.

Stephansson, O. (1981) The Näsliden project-rock mass investigations. In: Stephansson, O. & Jone, M. J. (eds.) *Proceedings of Application of Rock Mechanics to Cut and Fill Mining*. Institution of Mining and Metallurgy, London. pp. 145–161.

Stephansson, O. & Zang, A. (2012) ISRM suggested methods for rock stress estimation, Part 5: Establishing a model for the in-situ stress at a given site. *Rock Mechanics and Rock Engineering*, 45, 955–969.

Stille, H. & Palmström, A. (2003) Classification as a tool in rock engineering. *Tunnelling and Underground Space Technology*, 18(4), 331–345.

Stille, H. & Palmström, A. (2008) Ground behaviour and rock mass composition in underground excavations. *Tunnelling and Underground Space Technology*, 23, 46–64.

Swan, G. (1983) Deformation of stiffness and other joint properties from roughness measurements. *Rock Mechanics and Rock Engineering*, 16, 19–38.

Swoboda, G., Mertz, W. & Beer, G. (1987) Rheological analysis of tunnel excavations by means of coupled finite element (FEM)-boundary element (BEM) analysis. *International Journal for Numerical and Analytical Methods in Geomechanics*, 11, 115–129.

Tan, W., Kulatilake, P.H.S.W. & Sun, H. (2014a) Influence of an inclined rock stratum on in-situ stress state in an open-pit mine. *Geotechnical and Geological Engineering*, 32(1), 31–42.

Tan, W., Kulatilake, P.H.S.W., Sun, H. & Sun, Z. (2014b) Effect of faults on in-situ stress state in an open pit mine. *Electronic Journal of Geotechnical Engineering*, 19, 9597–9629.

Tang, C. A. (1995) Numerical simulation of rock failure process. *Proceedings of 2nd Youth Symposium on Rock Mechanics and Rock Engineering in China, 6–8 July, Chengdu*. China Science and Technology Press, Beijing. pp. 1–8.

Terzaghi, K. (1946) Rock defects and loads on tunnel support. In: Proctor, R. V. & White, T. L. (eds.) *Rock Tunneling with Steel Supports*. Commercial Shearing and Stamping Co., Youngstown, OH. pp. 17–99.

Thompson, P. M. & Chandler, N. A. (2004) In situ rock stress determinations in deep boreholes at the Underground Research Laboratory. *International Journal of Rock Mechanics and Mining Sciences*, 41, 1305–1316.

Tsesarsky, M. & Hatzor, Y. H. (2006) Tunnel roof deflection in blocky rock masses as a function of joint spacing and friction: A parametric study using discontinuous deformation analysis (DDA). *Tunnelling and Underground Space Technology*, 21, 29–45.

Vardakos, S. S., Gutierrez, M. S. & Barton, N. R. (2007) Back-analysis of Shimizu tunnel no. 3 by distinct element modeling. *Tunnelling and Underground Space Technology*, 22, 401–413.

Vlachopoulos, N. & Diederichs, M. S. (2014) Appropriate uses and practical limitations of 2D numerical analysis of tunnels and tunnel support response. *Geotechnical and Geological Engineering*, 32, 469–488.

Vyazmensky, A., Elmo, D., Stead, D. & Rance, J. R. (2007) Combined finite-discrete element modelling of surface subsidence associated with block caving mining. In: Eberhardt, E., Stead, D. & Morrison, T. (eds.) *Rock Mechanics: Meeting Society's Challenges and Demands: Proceedings of the 1st Canada-U.S. Rock Mechanics Symposium, Vancouver, Canada*. Taylor & Francis Group, London. 1. pp. 467–475.

Wang, M., Kulatilake, P.H.S.W., Um, J. & Narvaiz, J. (2002) Estimation of REV size and three-dimensional hydraulic conductivity tensor for a fractured rock mass through a single well packer test and discrete fracture fluid flow modeling. *International Journal of Rock Mechanics and Mining Sciences*, 39(7), 887–904.

Wang, S., Zheng, H., Li, C. & Ge, X. (2011) A finite element implementation of strain-softening rock mass. *International Journal of Rock Mechanics and Mining Sciences*, 48, 67–76.

Wang, X., Kulatilake, P.H.S.W. & Song, W. D. (2012) Stability investigations around a mine tunnel through three-dimensional discontinuum and continuum stress analyses. *Tunnelling and Underground Space Technology*, 32, 98–112.

Wawersik, W. R. & Fairhurst, C. (1970) A study of brittle rock fracture in laboratory compression experiments. *International Journal of Rock Mechanics and Mining Sciences*, 7, 561–575.

Wickham, G. E., Tiedemann, H. R. & Skinner, E. H. (1972) Support determinations based on geological predictions. In: Lane, K. S. & Garfield, L. A. (eds.) *Proceedings of North American Rapid Excavation and Tunneling Conference, Chicago*. American Institute of Mining, Metallurgical, and Petroleum Engineers, New York. 1. pp. 43–64.

Wu, J. H., Juang, C. H. & Lin, H. M. (2005) Vertex-to-face contact searching algorithm for three-dimensional frictionless contact problems. *International Journal for Numerical Methods in Engineering*, 63(6), 876–897.

Wu, Q. & Kulatilake, P.H.S.W. (2012a) REV and its properties on fracture system and mechanical properties, and an orthotropic constitutive model for a jointed rock mass in a dam site in China. *Computers and Geotechnics*, 43, 124–142.

Wu, Q. & Kulatilake, P.H.S.W. (2012b) Application of equivalent continuum and discontinuum stress analyses in three-dimensions to investigate stability of a rock tunnel in a dam site in China. *Computers and Geotechnics*, 46, 48–68.

Xing, Y., Kulatilake, P.H.S.W. & Sandbak, L. A. (2016) Investigation of rock mass stability around tunnels in an underground mine in USA by 3-D numerical modeling. *Proceedings of the 35th International Conference on Ground Control in Mining, Morgantown, 26–28 July, West Virginia, USA*. Society for Mining, Metallurgy and Exploration. pp. 213–221.

Xing, Y., Kulatilake, P.H.S.W. & Sandbak, L. A. (2017) Rock mass stability investigation around tunnels in an underground mine in USA. *Geotechnical and Geological Engineering*, 35, 45–67.

Xing, Y., Kulatilake, P.H.S.W. & Sandbak, L. A. (2018a) Investigation of rock mass stability investigation around the tunnels in an underground mine in USA using three-dimensional numerical modeling. *Rock Mechanics and Rock Engineering*, 51, 579–597.

Xing, Y., Kulatilake, P.H.S.W. & Sandbak, L. A. (2018b) Effect of rock mass and discontinuity mechanical properties and delayed rock supporting on tunnel stability in an underground mine. *Engineering Geology*, 238, 62–75.

Xing, Y., Kulatilake, P.H.S.W. & Sandbak, L. A. (2019) Rock mass stability investigation around tunnels in an underground mine in USA. *International Journal of Geomechanics*, 19(5), 05019004.

Xu, N., Kulatilake, P.H.S.W., Tian, H., Wu, X., Nan, Y. & Wei, T. (2013) Surface subsidence prediction for the WUTONG mine using a 3-D finite difference method. *Computers and Geotechnics*, 48, 134–145.

Yazdani, M., Sharifzadeh, M., Kamrani, K. & Ghorbani, M. (2012) Displacement-based numerical back analysis for estimation of rock mass parameters in Siah Bisheh powerhouse cavern using continuum and discontinuum approach. *Tunnelling and Underground Space Technology*, 28, 41–48.

Yeung, M. R. & Leong, L. L. (1997) Effects of joint attributes on tunnel stability. *International Journal of Rock Mechanics and Mining Sciences*, 34, 3–4.

Yeung, M. R., Jiang, Q. H. & Sun, N. (2007) A model of edge-to-edge contact for three-dimensional discontinuous deformation analysis. *Computers and Geotechnics*, 34, 175–186.

Yeung, M. R., Sun, N., and Jiang, Q. H. (2008) Tunnel stability analysis using a method coupling block theory and 3D DDA. *Proceedings of the 42nd U.S. Rock Mechanics Symposium, 29 June–2 July, San Francisco, California*. American Rock Mechanics Association. pp. 927–932.

Yoshida, H. & Horii, H. (1998) Micro-mechanics based continuum analysis for the excavation of large-scale underground cavern. *Proceedings of SPE/ISRM Rock Mechanics in Petroleum Engineering, 8–10 July, Trondheim, Norway*. Society of Petroleum Engineers. 1. pp. 209–218.

Zang, A., Stephansson, O., Heidbach, O. & Janouschkowetz, S. (2012) World stress map database as a resource for rock mechanics and rock engineering. *Geotechnical and Geological Engineering*, 30, 625–646.

Zhao, X. G. & Cai, M. (2010) Influence of plastic shear strain and confinement-dependent rock dilation on rock failure and displacement near an excavation boundary. *International Journal of Rock Mechanics and Mining Sciences*, 47, 723–738.

Zhang, G., He, F. & Jiang, L. (2016) Analytical analysis and field observation of break line in the main roof over the goaf edge of longwall coal mines. *Mathematical Problems in Engineering*, (9), 1–11.

Zhang, L. (2016) Determination and applications of rock quality designation (RQD). *Journal of Rock Mechanics and Geotechnical Engineering*, 8(3), 389–397.

Zhang, L. & Einstein, H. H. (2004) Using RQD to estimate the deformation modulus of rock masses. *International Journal of Rock Mechanics and Mining Sciences*, 41(2), 337–341.

Zhang, Q., Jiang, B. S., Wang, S., Ge, X. & Zhang, H. (2012) Elasto-plastic analysis of a circular opening in strain-softening rock mass. *International Journal of Rock Mechanics and Mining Sciences*, 50, 38–46.

Zhu, H., Wei, W., Chen, J., Ma, G., Liu, X. & Zhuang, X. (2016) Integration of three-dimensional discontinuous deformation analysis (DDA) with binocular photogrammetry for stability analysis of tunnels in blocky rockmass. *Tunnelling and Underground Space Technology*, 51, 30–40.

Zhu, W., Li, X., Guo, Y., Sun, A. & Sui, B. (2004) Systematical study on stability of large underground houses. *Chinese Journal of Rock Mechanics and Engineering*, 23(10), 1689–1693.

Zienkiewicz, O. C., Best, B., Dullage, C. & Stagg, K. (1970) Analysis of non-linear problems in rock mechanics with particular reference to jointed rock systems. *Proceedings of the Second International Congress on Rock Mechanics, Belgrade, Yugoslavia*. ISRM. 3. pp. 501–509.

Index